Stoke-on-Trent Libraries
Approved for Sale

STAFFORDSHIRE
COUNTY REFERENCE
LIBRARY
HANLEY
STOKE-ON-TRENT

ADVANCED TECHNICAL CERAMICS

ADVANCED TECHNICAL CERAMICS

Edited by
Shigeyuki Sōmiya
Research Laboratory of Engineering Materials
Tokyo Institute of Technology

738.134

Academic Press, Inc.
Harcourt Brace Jovanovich, Publishers
Tokyo   San Diego   New York   Berkeley
Boston   London   Sydney   Toronto

Copyright © 1984 by Shigeyuki Sōmiya
English translation copyright © 1989 by Academic Press Japan, Inc.
All Rights Reserved.
No part of this publication may be reproduced or transmitted in any form or by any means, electronic or mechanical, including photocopy, recording, or any information storage and retrieval system, without permission in writing from the publisher.

原著日本語版「© 1984 宗宮重行」は講談社が出版したものであり, 本英語版はアカデミック・プレス・ジャパンが講談社との契約により出版するものである。

Academic Press Japan, Inc.
Ichibancho Central Building, 22-1 Ichibancho
Chiyoda-ku, Tokyo 102, Japan

United States Edition published by
ACADEMIC PRESS, INC.
San Diego, California 92101

United Kingdom Edition published by
Academic Press Limited
24–28 Oval Road, London NW1 7DX

Library of Congress Cataloging-in-Publication Data

Advanced technical ceramics.

   Rev. translation from Japanese.
   Includes index.
   1. Ceramics. I. Sōmiya, Shigeyuki.
TP807.A33 1988   666   88-10456
ISBN 0-12-654630-4 (hardcover)(alk. paper)

Printed in the United States of America
89 90 91 92   9 8 7 6 5 4 3 2 1

CONTENTS

Contributors ix
Preface xi

Part I
Introduction to Ceramics

1 **Ceramics: Definitions**
   Shigeyuki Sōmiya
   I. Introduction   3
   II. Definitions   5
       References   9

2 **Types of Ceramics**
   Shigeyuki Sōmiya
   I. Introduction   11
   II. Classification by Chemical Composition   11
   III. Classification by Minerals   12
   IV. Classification by Molding Technique   12
   V. Additional Ways to Classify Ceramics   16
       References   16

3 **Synthetic Raw Materials for Ceramics**
   Akira Nagai and Yoshitaka Kimura
   I. Introduction   27
   II. General Issues Concerning Powders for Ceramics   29
   III. Major Materials for Ceramics   31
        References   63

4 **Production Processes for Ceramics**
   Isamu Fukuura and Tadayoshi Hirao
   I. Introduction   65
   II. Preparing the Body   66
   III. The Molding Process   69

v

vi Contents

        IV. The Firing Process   75
        V. Issues for the Near Future in Ceramics Production
           Techniques   81
           References   81

5 Evaluating Ceramics
Hiroshi Okuda

        I. Introduction   83
        II. Mechanical Properties   83
        III. Thermal Properties   93
           References   99

Part II
Properties and Applications of Ceramics

6 Electrical and Electronic Properties
Kikuo Wakino

        I. Introduction   103
        II. Insulating Properties   104
        III. Semiconductors   109
        IV. High-Conductivity Ceramics   114
        V. Superconductors   115
        VI. Ionic Conduction   118
           References   122

7 Magnetic Properties
Teitaro Hiraga

        I. Introduction   125
        II. Ferrite: An Oxide Magnetic Material   126
        III. Summary of Characteristics   127
        IV. Soft Magnetic Ferrite   129
        V. Hard Ferrite   134
        VI. Semihard Magnetic Ferrite   136
        VII. Ferrite for Microwave Use   139
           References   142

8 Thermal Properties
Tatsuyuki Kawakubo and Noboru Yamamoto

        I. Theory of Thermal Properties   145
        II. Ceramics That Exploit Thermal Properties   155
           References   165

Contents **vii**

9 Chemical Properties
Shigeru Hayakawa and Satoshi Sekido
- I. Introduction   167
- II. Chemical Sensors   168
- III. Chemical Batteries and Electric Double-Layer Capacitors   176
- IV. Chemical Pumps   179
- V. Electrochromism   180
- VI. High-Temperature Steam Electrolysis   181
- VII. Catalysts   183
  - References   184

10 Optical Properties
Toru Kishii and Hitoshi Hirano
- I. Noncrystalline Substances   189
- II. Crystals   199
  - References   207

11 Biological Applications
Kazuo Inamori
- I. Metallic Implants   209
- II. Biological Practice in Optimizing Implants   211
- III. Carbon and Ceramic Systems   211
- IV. Medical Applications   214
- V. Conclusion   221
  - References   221

12 Mechanical Properties
Shigetomo Nunomura, Junn Nakayama, Hiroshi Abe, Osami Kamigaito, Kitao Takahara, and Katsutoshi Matsusue
- I. Strength of Materials   223
- II. Mechanical Properties of Today's Ceramics   233
- III. Ceramic Parts for Automobiles   241
- IV. Engines   248
  - References   256

Part III
Machining Methods

13 Precision Machining Methods for Ceramics
Akira Kobayashi
- I. Introduction   261
- II. Cutting   264

III. Grinding 269
IV. Lapping and Polishing 279
V. Laser Processing 285
VI. Other Machining Methods 300
References 308

***Appendix: Chronology of the Development of Advanced Ceramics* 315**
Suezo Sugaike

Index 345

CONTRIBUTORS

Numbers in parentheses indicate the pages on which the authors' contributions begin.

Hiroshi Abe (223), Asahi Glass Co., Ltd., Kanagawa-ku, Yokohama 221, Japan
Isamu Fukuura (65), NGK Spark Plug Co., Ltd., Mizuho-ku, Nagoya 467, Japan
Shigeru Hayakawa (167), Matsushita Research Institute Tokyo, Inc., Tama-ku, Kawasaki 214, Japan
Teitaro Hiraga (125), TDK Corporation, Chuo-ku, Tokyo 103, Japan
Hitoshi Hirano (189), Toshiba Corporation, Saiwai-ku, Kawasaki 210, Japan
Tadayoshi Hirao (65), NGK Spark Plug Co., Ltd., Mizuho-ku, Nagoya 467, Japan
Kazuo Inamori (209), Kyocera Corporation, Yamashina-ku, Kyoto 607, Japan
Osami Kamigaito (223), Toyota Central Research & Development Laboratories, Inc., Aichi-gun, Aichi 480-11, Japan
Tatsuyuki Kawakubo (145), Tokyo Institute of Technology, Meguro-ku, Tokyo 152, Japan
Yoshitaka Kimura (27), Showa Denko, K.K., Minato-ku, Tokyo 105, Japan
Toru Kishii (189), Toshiba Glass Co., Ltd., Haibara-gun, Shizuoka 421-03, Japan
Akira Kobayashi (261), Ibaragi Polytechnic College, Mito 310, Japan
Katsutoshi Matsusue (223), National Aerospace Laboratory, Chofu, Tokyo 182, Japan
Akira Nagai (27), Showa Denko, K.K., Minato-ku, Tokyo 105, Japan
Junn Nakayama (223), Asahi Glass Co., Ltd., Kanagawa-ku, Yokohama 221, Japan
Shigetomo Nunomura (223), Tokyo Institute of Technology, Midori-ku, Yokohama 227, Japan
Hiroshi Okuda (83), Japan Fine Ceramics Center, Atsuta-ku, Nagoya 456, Japan
Satoshi Sekido (167), Matsushita Research Institute Tokyo, Inc., Tama-ku, Kawasaki 214, Japan
Shigeyuki Sōmiya[1] (3, 11), Research Laboratory of Engineering Materials, Tokyo Institute of Technology, Midori-ku, Yokohama 227, Japan

---

[1]Present address: The Nishi Tokyo University, 3-7-19 Seijo, Setagaya, Tokyo 157, Japan.

Suezo Sugaike (315), Sci-Tech Research Co., Ltd., Chiyoda-ku, Tokyo 100, Japan

Kitao Takahara (223), National Aerospace Laboratory, Chofu, Tokyo 182, Japan

Kikuo Wakino (103), Murata Manufacturing Company Limited, Nagaoka-kyo-city 617, Japan

Noboru Yamamoto (145), NGK Insulators Co., Ltd., Mizuho-ku, Nagoya 467, Japan

PREFACE

Advanced Technical Ceramics was originally published in Japanese to address recent developments in technical ceramics, especially in Japan. It was written for engineers, scientists, students, and others interested in new fields in ceramics. The Japanese edition was so successful that Academic Press approached me about preparing a translation for English-language readers.

Advanced Technical Ceramics gives a thorough overview of technical ceramics. It covers all aspects of technical ceramics: definitions, discussion of raw materials, electronic and mechanical materials and processes, and biomaterials. The contributors, all experts in their fields, have updated their articles for this English edition and have reviewed the translation for accuracy. This edition is more than a translation, as many papers have been improved as well as brought up to date.

I wish to express my appreciation to Mr. Junichiro Minagawa of Academic Press and to the translator, Ms. Ruth S. McCreery. It would have been impossible to publish this book without their efforts.

It is my hope that *Advanced Technical Ceramics* will promote the understanding and development of ceramics around the world.

Shigeyuki Sōmiya

I
INTRODUCTION TO CERAMICS

1

Ceramics: Definitions

Shigeyuki Sōmiya
Research Laboratory of Engineering
  Materials
Tokyo Institute of Technology
Midori-ku, Yokohama 227, Japan

I. INTRODUCTION

The first ceramics were the low-firing earthenwares that appeared some 10,000 years ago when humans had mastered the use of fire. Those early potters used simple pit firing, for they lacked kilns (specially designed ovens) for firing their wares. Nonetheless, the production of those early, unsophisticated earthenwares was the starting point for the subsequent development of ceramics.

Until perhaps 100 years ago, the word *ceramics* meant pottery; ceramic products were limited to tableware for everyday use, roofing tiles, clay pipe, and brick.

The mass production of iron and steel, which began in the latter half of the nineteenth century, required the development of new refractory ceramics, entailing a shift from the use of conventional fireclays to refractory materials containing silica, alumina, chrome-magnesia, or magnesia. Without these advances in refractory materials, the production of steel and other metals could not have attained its present development.

Ceramics have made essential contributions in many other sectors of industrial society as well. For instance, the rapid production of textiles

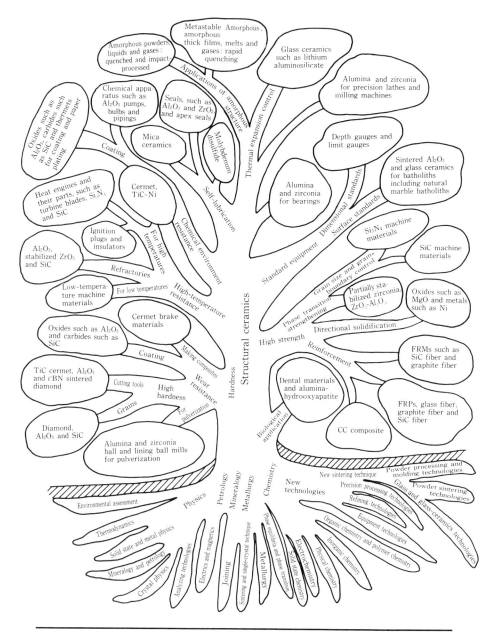

Figure 1.1. Structure of basic and applied ceramics. Adapted from F. C. Report (1983).

made of synthetic fibers has been made possible by the advent of thread guides of alumina and other ceramic materials. Bisque wares are used in biotechnology and in sewage disposal. They are just as much ceramic products as are teacups or flower pots for bonsai.

Today, ceramic products touch our lives in many ways, as shown in Figure 1.1. Pottery and porcelain vessels, glass, and cement are only among the more familiar. Their applications include the magnets in television sets, optical fibers for telecommunications, automobile spark plugs, and the insulators for Japan's high-speed trains. They are widely used in electronics, not only as magnets and as insulators, but also as heating elements and substrates for integrated circuits. As engineering ceramics, they appear in ceramic engines and cutting tools. In bioceramics, they are used for artificial teeth and bones.

II. DEFINITIONS

While ceramics are finding applications in many fields, agreement is not complete on the precise meaning of the term.

> Shiraki (1963, p. 304) has written that ceramics refers to non-metallic products that, having undergone processing at high temperatures, have been given the distinctive features of ceramics; [ceramics] may include non-metallic, inorganic products which have received processing at high temperatures in the course of their manufacture.

If a more precise definition is desired, however, it is necessary to consult additional sources. W. W. Perkins, editor of the American Ceramic Society's *Ceramic Glossary* (1984, pp. 13–14) defines *ceramics* as follows:

> ceramic sing. n.; ceramics pl. n. (1) A general term applied to the art or technique of producing articles by a ceramic process, or to the articles so produced. C-242 (2) Any of a class of inorganic, nonmetallic products which are subjected to a high temperature during manufacture or use. (High temperature usually means a temperature above a barely visible red, approximately 540°C (1000°F)). Typically, but not exclusively, a ceramic is a metallic oxide, boride, carbide, or nitride, or a mixture or compound of such materials; that is, it includes anions that play important roles in atomic structures and properties.

> ceramic adj. (1) Of or pertaining to ceramics, that is, inorganic, nonmetallic as opposed to organic or metallic. (2) Pertaining to products manufactured from inorganic nonmetallic substances which are subjected to a high temperature during manufacture or use. (3) Pertaining to the manufacture or use of such articles or materials, such as ceramic process or ceramic science.

Loran S. O' Bannon, author of the *Dictionary of Ceramic Science and Engineering* (1983, p. 54) offers this definition:

> ceramic. (1) Any of a class of inorganic, nonmetallic products which are subjected to a temperature of 540°C (1000°F) and above during manufacture or use, including metallic oxides, borides, carbides, or nitrides, and mixtures or compounds of such materials.
> (2) Pertaining to ceramics.
> (3) Pertaining to the manufacture or use of ceramic processes, articles, materials, technology, and science.

The two preceding definitions are American. In British usage, according to D. E. Dodd's *Dictionary of Ceramics* (1964, p. 60), the following definition applies:

> Ceramic. The usual derivation is from *Keramos*, the Greek work for potters' clay or ware made from clay and fired; by a natural extension of meaning, the term has for long embraced all products made from fired clay, i.e. bricks and tiles, pipes and fireclay refractories, sanitary-ware and electrical porcelain, as well as pottery tableware. In 1822 silica refractories were first made; they contained no clay, but were made by the normal ceramic process of shaping a moist batch, drying the shaped ware and firing it. The word 'ceramic', while retaining its original sense of a product made from clay, thus tacitly began to include other products made by the same general process of manufacture. There has in consequence been no difficulty in permitting the term to embrace the many new non-clay materials now being used in electrical, nuclear and high-temperature engineering.
> In the USA a radical extension of meaning was authorized by the American Ceramic Society in 1920; chemically, clay is a silicate and it was proposed that the term 'ceramic' should be applied to all the silicate industries; this brought in glass, vitreous enamel, and hydraulic cement. In Europe, this wider meaning of the word has not yet been fully accepted.

These definitions reveal basic inconsistencies in what can be called ceramics. In the British definition, for instance, glass is not included among ceramics, but the broader U.S. and Japanese versions do admit glass to the ceramics family.

Technological change is also forcing changes in the limits placed on the term ceramics. In the past, the use of *ceramics* referred to technology, science, or art relating to nonmetallic, inorganic solid materials or to the production of goods from them or to their use. This definition covers all ceramic products, which may take many forms: polycrystalline material, single crystals, and amorphous materials in bulk, lumps, grains, thick or thin films, and fibers.

Crystalline ceramics include long-familiar ceramic products: pottery and porcelain vessels, refractories, glass, and cement. It more broadly includes electronic ceramics, including magnetic substances, insulators, integrated circuit substrates, dielectric substances, and heating elements; engineering ceramics, including ceramic engines and cutting tools; and bioceramics, including ceramic teeth and bones. These ceramic products in general are inorganic, nonmetallic solid materials formed at high temperatures from crystalline or amorphous materials, pores, and liquids.

The ceramics industry traces its roots back to those early earthenwares at the beginning of human history. Over the ages, humankind has developed a range of products formed of bodies based on natural materials such as clay and silicates—dishes and other tableware for daily use of pottery and porcelain, refractory substances, cement, and bottle glass. These are known as classical, traditional, or conventional ceramics. In contrast, ceramics formed of bodies with non-naturally occurring materials such as alumina, zirconia, or titania, or of synthetic materials, are called *new ceramics* or *modern ceramics*.

New ceramics was a term used quite frequently, particularly in Great Britain, in the 1940s and 1950s. After a period of years, however, the new ceramics could not hope to retain their newness. In addition, the development of ceramics in connection with nuclear power gave rise to new materials such as carbides and nitrides, called *special ceramics*. The first special ceramics conference was held in the early 1960s and conferences continue to be held to this day.

New ceramics, modern ceramics, and special ceramics embraced ceramics containing carbides, nitrides, and intermetallic compounds. The terms *special refractories* and *special pottery and porcelains* included pure oxide ceramics and ceramics for chemistry laboratory and electronic engineering use.

The term *fine ceramics* has come into use recently. The term refers to the very small particulate materials that are used to form the desired shapes and to the small grains that form the sintered product. However, fine ceramics, in the product called fine ceramic bodies, has been a widely used term since the 1800s (Shiraki, 1976, p. 1065). In Europe and the United States, fine-grained ceramic products are called fine ceramics, as Norton's (1978) work indicates.

The term *fine ceramic* (*fain seramikkusu*) is also used in Japan. Fine ceramics, in the Japanese sense, are most similar to what are called advanced ceramics or technical ceramics in Europe and America. With the growth of Japan's economic power, there have been cases in which in France, for instance, ceramists have adopted the term fine ceramics as it is used in Japan.

In Japan, according to the Japanese Ministry of International Trade and Industry (MITI); (Japan 1985), in contrast to conventional ceramics, "Fine ceramics are high value added inorganic materials produced from high purity synthetic powders to control microstructure and properties."

Technical ceramics is a term that includes all industrial ceramics except pottery or porcelain tablewares, home plumbing fixtures, and the like. It covers electronic ceramics such as dielectrics and insulators, ceramic magnets for television and other applications, structural ceramics used for engine parts, and all other industrial applications.

Technical ceramics are defined as those ceramics that exhibit a high degree of industrial efficiency through their carefully designed microstructures and superb dimensional precision. In technical ceramics, rigorously selected materials are used in products with a precisely regulated chemical composition, fabricated under strictly controlled methods of shaping and firing.

Engineering ceramics is a technical term sometimes used in the same way as technical ceramics. In general, technical ceramics has a broader meaning; engineering ceramics is often more strictly confined to ceramics for structural applications. Thus, engineering ceramics is a subset of technical ceramics, including principally ceramics with superior mechanical properties, such as great strength, abrasion resistance, a high level of elasticity, enhanced hardness, heat resistance, and lubricating ability. These ceramics have found applications in ceramic engines, cutting tools, grinding materials, and materials for bearings.

Advanced ceramics or *high-technology ceramics* is used for many of the same materials and products as technical ceramics. These terms, however, particularly emphasize the special value or advanced features which heighten the commercial value of technical ceramics. Like engineering ceramics and technical ceramics, they are general designations for ceramics used in industrial applications.

Structural ceramics are used as structural parts in machines. They include engine parts, cutting tools, and seals. *Bioceramics* are, as the name indicates, ceramics used in biological applications, such as artificial teeth or bones. *Electroceramics* are suitable for use as insulators for electric line or in electrical components. *Electronic ceramics*, however, are ceramics used in the electronics field, including dielectric substances, magnetic substances, and semiconductors.

Hydrothermal ceramics are powders, single crystals, sintered substances, and thin membranes formed in a vessel resistant to high temperatures and pressures (an autoclave, for instance) under high-pressure and high-temperature conditions with pure water, aqueous solutions, organic solvents, and/or nonaqueous solvents. The powders, single crystals, sintered substances, and thin films used under those conditions are also called hydrothermal ceramics.

Recently, the use of polymers such as SiC as precursor materials for ceramics has led to a strong demand for high-performance ceramics. Because of their enhanced features, these ceramics are called *high-performance ceramics, active ceramics* or *high-value-added ceramics*. J. A. Pask and H. Kent Bowen (personal communication, 1980) have argued that active ceramics or high-value-added ceramics would be more suitable terms than high-performance ceramics. Those high-performance features of ceramics will be discussed in detail later in this book.

In the history of ceramics, reflected in the host of terms discussed here, the traditional conception of ceramic products was based on the assumption that ceramics were container and partition materials, with the distinctive features of hardness, incombustibility, high corrosion resistance, and fragility. Today's ceramics industry is shifting, however, to a search for new properties of ceramic materials and to the creation of new materials making use of these properties.

The term *chemically bonded ceramics* (CBC) designates a group of ceramic materials in which the consolidation process involves only chemical reactions which occur at low temperatures. Thus CBCs are distinguished from traditional ceramics by two parameters: (1) their bonding is not achieved by thermally stimulated diffusion and (2) bonding is achieved at or near room temperature. The best known examples are the many kind of cements, and in nature, bones, teeth, and invertebrate skeletons (R. Roy, personal communication, 1980).

REFERENCES

Dodd, D. E. (1964). *Dictionary of Ceramics*. Amsterdam: Elsevier.

F. C. *Report* (1983). 11, May.

Japan, Ministry of International Trade and Industry. (1985). Tsushō sangyō shō. Fain seramikkusu shitsu [Advanced Ceramics Division, MIT I]. *Nikkei New Materials*, 11–11, p. 72. Tokyo: Nikkei McGraw-Hill.

Norton, F. H. (1978). *Fine Ceramics*. Melbourne, FL: Krieger.

O'Bannon, Loran S. (1984). *Dictionary of Ceramic Science and Engineering*. New York: Plenum.

Perkins, W. W. (ed.) (1984). *Ceramic Glossary*. Columbus, OH: The American Ceramic Society.

Shiraki, Yōichi (1963). *Yōgyō jiten* [Ceramics dictionary]. (Yōgyō kyōkai, ed.) Tokyo: Maruzen.

Shiraki, Yōichi. (1976). *Fain seramikkusu* [Fine ceramics]. Tokyo: Gihōdō.

2
Types of Ceramics

Shigeyuki Sōmiya
Research Laboratory of Engineering
 Materials
Tokyo Institute of Technology
Midori-ku, Yokohama 277, Japan

I. INTRODUCTION

There are many possible approaches to classifying ceramics. They can be grouped, for instance, by their chemical composition, their mineral content, the processing methods used in their production, their properties, or their uses. Their properties and uses will be discussed in detail in Part II. Therefore, this chapter provides an exposition of ways that ceramics can be classified by chemical composition, mineral content, and processing methods used in production, summarizing the results in tables and diagrams as much as possible.

II. CLASSIFICATION BY CHEMICAL COMPOSITION

Ceramics contain a great variety of oxides, of which ternary and quaternary compounds are most common (Tabe, Seiyama, and Fueki, 1978; Yoshiki, 1967). In addition, carbides and nitrides are also found in ceramics. Table 2.1 summarizes the classification of ceramics by their chemical composition.

Table 2.1 One Possible Classification of Ceramics by Chemical Composition

Oxides	
Binary compounds	SiO_2, Al_2O_3, Fe_2O_3, FeO, Fe_3O_4, CaO, MgO, Mn_3O_4, TiO_2, ZrO_2, HfO_2, ThO_2, BeO, Y_2O_3, La_2O_3, CeO_2
Ternary compounds	$3Al_2O_3 \cdot 2SiO_2$-$2Al_2O_3 \cdot SiO_2$, Al_2SiO_5, $MgSiO_3$, Mg_2SiO_4, $CaO \cdot MgO$, Ca_2SiO_3, $Ca_3Si_2O_7$, $MgAl_2O_4$, $FeCr_2O_4$, $MgFe_2O_4$, $MgCr_2O_4$, $FeTiO_3$, $CaTiO_3$, $3Y_2O_3$, $5Fe_2O_2(Y_6Fe_{10}O_{24})$, $BaO \cdot 6Fe_2O_3(BaFe_{12}O_{19})$, Ba_2SiO_4(phenacite)
Quaternary compounds	$3CaO \cdot MgO \cdot 2SiO_2$(merwinite), $2CaO \cdot MgO \cdot 2SiO_s$(akermanite), $3BeO \cdot Al_2O_2$, $6SiO_2$(beryl), $2MgO \cdot 2Al_2O_3 \cdot 5SiO_2$(cordierite), $Li_2O \cdot Al_2O_3 \cdot 2SiO_2$(eucryptite), $2CaO \cdot Al_2O_3 \cdot SiO_2$(gehlenite), $3CaO \cdot Al_2O_3 \cdot 2SiO_2$(grossuralite), $Na_2O \cdot Al_2O_3 \cdot SiO_2$ (nephelite), $3MgO \cdot Al_2O_3 \cdot 3SiO_2$ (pyrope), $Li_2O \cdot Al_2O_3 \cdot 4SiO_2$(spodumene)
Carbides	SiC, WC, TiC, TaC, ZrC, B_4C
Nitrides	Si_3N_4, BN, TiN, ZrN, AlN
Oxynitrides	Sialon

III. CLASSIFICATION BY MINERALS

A detailed explanation of the minerals in ceramics is beyond the scope of this book. Therefore, we limit ourselves to the following list of minerals that occur in ceramics: corundum (Al_2O_3), mullite ($3Al_2O_3 \cdot 2SiO_2 - 2Al_2O_3 \cdot SiO_2$), cristobalite, tridymite, quartz, condiertie, yttrium iron garnet, apatite, akermanite, anorthite, phenacite, perovskite, spinel, and zircon.

IV. CLASSIFICATION BY MOLDING TECHNIQUE

Ceramics may be molded by hand, by die pressing (mechanical pressing, dry pressing, die molding), by isostatic pressing (rubber pressing, isostatic pressing, and hydrostatic pressing), by casting (slip casting, vibratory casting, solid casting), injection molding, extrusion, and tape formation (by the doctor blade or calendar roll method or by extrusion). The most important of these methods are discussed below.

A. ISOSTATIC PRESSING

Die pressing is a simple method, easily applied, but it has an invariant disadvantage: It is difficult to apply uniform pressure on the particles of the ceramic body being molded. The unequal pressures result in strains, which produce qualitative differences in the products. Quality tends to be quite inconsistent. Isostatic pressing, in contrast, increases the number of points of contact between the particles, facilitating the progress of the reaction. In addition, this method minimizes frictional resistance with the wall surfaces. Methods which apply pressure evenly are also called rubber pressing or hydrostatic pressing. The preliminary molding is carried out at less than 30 MPa. This molded form is then inserted into a thin bag of rubber or other material, where, using a medium such as water or glycerin, it is press-molded at 50 to 400 MPa. An alternative method is to pack particles into a rubber mold and press them by passing the mold through a liquid.

In Europe and the United States, these methods have become established as casting methods in both the laboratory and industry. They have the advantages of (1) reducing the reaction time for solidification, (2) permitting the production of bodies in a range of complex shapes, and (3) reducing variations in the physical properties of the ceramics thus formed.

B. SLIP CASTING

A fine powder of the ceramic materials in solution with 30% or more of water or other solvent is made into a slip in a ball mill. The slip is poured into a mold (made of plaster of paris if the solvent is water; with other solvents, filter paper is used for absorption). After a set period of time, the excess slip is poured from the mold, and the object is removed. This method yields molded objects of a fixed thickness. The distinctive features of slip casting are the delicacy and accuracy of the objects that can be cast in this way, the high rate of contraction, and the considerable time required for drying. Sometimes water glass or fine grains of deflocculants are added when the slip is made.

C. VIBRATORY SOLID CASTING

Powder and water are mixed, with water forming 10% or less. Vibrations applied to the powder in the mold create vibrations in it that shape the body. This casting method has a lower rate of shrinkage than does slip casting, the drying stage is simplified, and a plaster of paris mold is not needed. Vibratory equipment, however, is needed. Solid casting can yield delicately and accurately molded objects.

D. EXTRUSION AND INJECTION MOLDING

In extrusion, a kneaded body is forced out of a nozzle at room temperature. In injection molding, a thermoplastic bonding agent is added, the

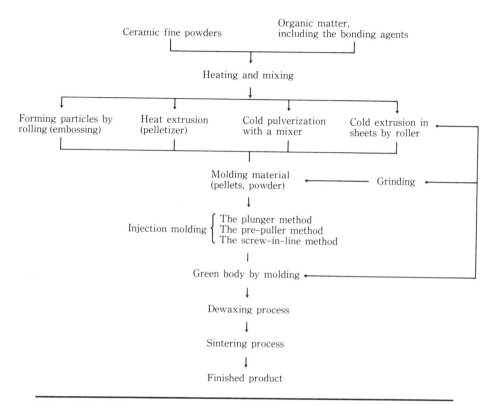

Figure 2.1. The injection molding process. Adapted from Saitō (1983).

mixture is kneaded, and then heated to a suitable temperature and forced out of a nozzle under pressure. These methods are used particularly in forming ceramic objects that are longer than they are wide, such as sticks or pipes. A bonding agent is added to the powdered ceramic body, making it sticky. The mixture is kneaded well and extruded or injected through a nozzle of the appropriate shape. Since the added bonding agent often is moree than 10% of the mixture, conformation to the dissolution of the bonding agent under heat is necessary in drying and heating the molded object. Figure 2.1 presents the injection of molding process.

E. THE DOCTOR BLADE METHOD

This method produces green tape by painting a determined thickness of a slip composed of the mixed basic materials with a doctor blade, evaporat-

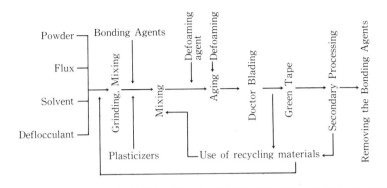

Figure 2.2. Simplified diagram of the doctor blade process. Adapted from Saitō (1983).

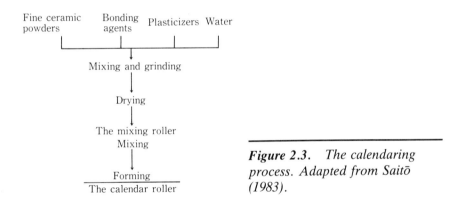

Figure 2.3. The calendaring process. Adapted from Saitō (1983).

ing off the solvent in drying the tape, and solidifying the ceramic. Figure 2.2 illustrates the process.

F. THE CALENDAR ROLL METHOD

A slip composed of the mixed basic materials is flaked while being dried on a rotary dryer. Then the mixing roll mixes it into homogeneity as it is heated. The resulting mixture is processed into a thin film by the debubbling, rolling, and finishing rollers. Figure 2.3 shows the calendaring process and Figure 2.4 the types of rollers.

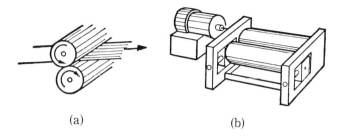

Figure 2.4. *Types of rollers. (a) shows the calendar roller and (b) the mixing roller. Adapted from Saitō (1983).*

V. ADDITIONAL WAYS TO CLASSIFY CERAMICS

Classification by methods of sintering and heating, their advantages and disadvantages, and their applications are summarized in Table 2.2. Figure 2.5 and 2.6 provide two alternative classifications of ceramics by their production technologies. For a classification of ceramics by their properties, see Figure 2.7. Table 2.3 presents a breakdown of ceramic types by their properties and uses.

REFERENCES

Ho, Taishin. (1982). 80 nendai no tenkai ga kitaisareru fain seramikkusu no tokusei to yōto [Characteristics and uses of fine ceramics, for which developments in the 1980s are promising]. Shūkan tōyō keizai, May 15, p. 79.

Kōgyō Gijutsuin, Gijutsu Shinkōka (ed.). (1981). *Kakushin gijutsu e no chosen (Gijutsu rikkoku o mezashite)* [The challenge of innovative technology (To establish a state on the basis of technology)]. Tokyo: Kōgyō Gijutsuin.

Saitō, Katsuyoshi. (1983). *Enjiniaringu seramikkusu* [Engineering ceramics]. *CMC R&D Repōto* [CMC R&D report], No. 37, pp. 39–79.

Tabe, Kozo; Seiyama, Tetsurō; and Fueki, Kazuo (eds.) *Kinzoku Sankabutsu to fukugō sankabutsu* [Metallic oxides and compound oxides]. Tokyo: Kōdansha, 1978.

Taki, Sadao. (1976). "Nyū seramikkusu no tsukaikata" [Uses of the new ceramics]. In Nyū seramikkusu konwakai (ed.), Tan kessho no ikusei [Growth of single crystals]. Osaka: Nyū seramikkusu konwakai.

Yoshiki, Bunpei. *Kōbutsu kōgaku* [Mineral technology]. Tokyo: Gihōdō, 1967.

Table 2.2. Classification by Method of Heating or Sintering

Processing Method[a]	Procedure	Advantages	Disadvantages	Examples
Pressureless sintering	Traditional method	1. Can produce products in complex shapes. 2. Can be used in mass production.	1. Shrinkage. 2. Product has small pores. 3. Strength at times is slightly inferior.	Al_2O_3, MgO, ZrO_2
Hot pressing	1. Body is placed in mold as powder, high temperature and high pressure are applied simultaneously, and the product is formed and sintered. 2. The mold is made of carbon, alumina, SiC, etc.	1. Little grain growth. 2. Can yield high-density products. 3. Can sinter ceramic powders normally difficult to sinter.	1. Mold is required. 2. There is a size and shape limitation. 3. Warps in the direction of pressing.	Al_2O_3, MgO, CaF_2
Hot isostatic pressing	Powder is packed into a capsule which can withstand high temperatures and is heated and pressed by a gas at high temperature and pressure.	1. Can be used to produce high-strength products with few defects. 2. Joints readily.	1. Requires high-temperature and -pressure gas equipment. 2. Requires a capsule.	Al_2O_3, Si_3N_4, MgO, ZrO_2
Reaction sintering	Uses the solid-gas phase and solid-liquid phase reaction to synthesize the ceramic powder and simultaneously sinter it.	1. Can produce products with complex shapes. 2. In some cases, there is no shrinkage through sintering.	1. Product is porous in some cases. 2. Product is weak in some cases.	Si_3N_4

(continued)

Table 2.2. (*Continued*)

Processing Method[a]	Procedure	Advantages	Disadvantages	Examples
Liquid-phase sintering	At high temperatures, the liquid phase is formed and the sintering aid can work effectively.	Can yield a high-density body through relatively low-temperature sintering.	Since the liquid phase is formed at a high temperature, in some cases strength of product at high temperatures may be weakened.	1. Si_3N_4 $SiO_2 + MgO \rightarrow$ liquid phase. 2. CaO, SiO_2 in ferrite.
Very (ultra) high-pressure sintering	Uses very-high-pressure equipment for sintering.	Can produce high-density sintered materials.	1. Cannot be used to produce large-scale products. 2. Requires very-pressure equipment.	Diamond, cubic BN, Si_3N_4
Shock-waved sintering	Uses shock waves generated by explosives or by other methods to apply very high temperature and high pressure for a short interval.	Can sinter in a short time.	1. Requires special equipment. 2. Cannot be used to make products in complicated shapes or large-scale products.	Cubic BN
CVD sintering	Produces a thick film by CVD to form the sintered substance.	Generally results in high degree of purity.	1. Gas bubbles tend to be left in substances. 2. Products are often susceptible to corrosion. 3. Cannot be used to produce large-scale or thick-bodied products.	TiB_2
Hydrothermal sintering	Instead of gas at high temperature and pressure, water or another fluid is used.	1. Can produce sintered materials containing volatile materials. 2. Can produce composite materials composed of organic materials or of organic and inorganic materials.	Requires hydrothermal equipment.	Mica, hydroxyapatite

Method	Description	Characteristics	Examples
Hydrothermal reaction sintering	Synthesizes powder by chemical reaction with an aqueous solution (or a fluid) and then sinters it.	3. Low-temperature sintering. 1. Can achieve a sintered substance with a finer grain than the original materials. 2. Can yield fine-grained sintered body.	Cr_2O_3, ZrO_2, HfO_2, $LaCrO_3$, $LaFeO_3$
Post normal sintering	A. Crystallizes the liquid phase after sintering. B. Applies HIP after pressureless sintering.	3. Low-temperature sintering. 1. Product is finer, stronger. 2. Defects are reduced.	Requires hydrothermal equipment. 1. Costs are high. 2. Requires HIP equipment.	A. Si_3N_4–Y_2O_3–Al_2O_3 B. Spinel, Al_2O_3
Atmospheric sintering	Particular regulation during sintering of the atmosphere and pressure of oxygen, nitrogen, water vapor, carbon dioxide, or other gas.	1. Valency control by atmosphere is possible. 2. Control of Fe^{2+}, Fe^{3+}, etc. is possible.	Fe_3O_4, $ZnFe_2O_4$, Ferrite, Si_3N_4
Melting	Used with glass and other materials.	Yields dense glass. Requires melting.	Glass, lenses, cast refractories
Crystallization	Melted like glass and then crystallized by heating after forming.	Yields thermal shock resistant porcelain vessels. Requires melting at high temperatures and reheating for crystallization.	Pyroceram

^a In addition to the processing methods listed in the table, the following are also used in producing ceramics: hot extrusion, hot rolling, hot injection, hot forging, hot extension, thermal etching, hot cutting, fused cast processing, electrofused cast processing, flame spray coating processing, and flamegun-formed block processing.

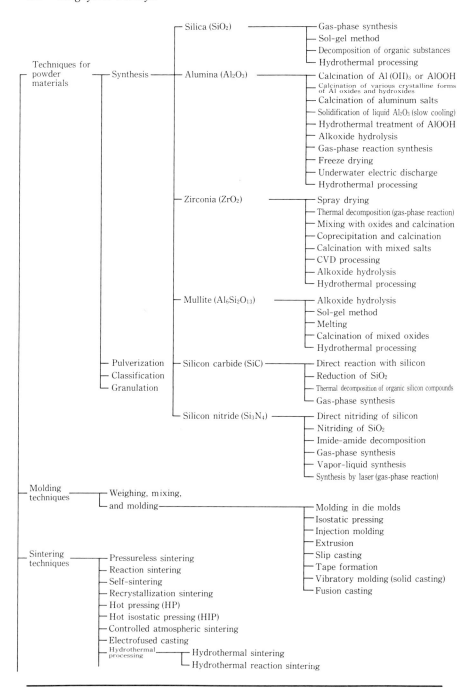

Figure 2.5. Processing of advanced ceramics. Adapted from Kōgyō Gijutsuin (1981).

2. Types of Ceramics

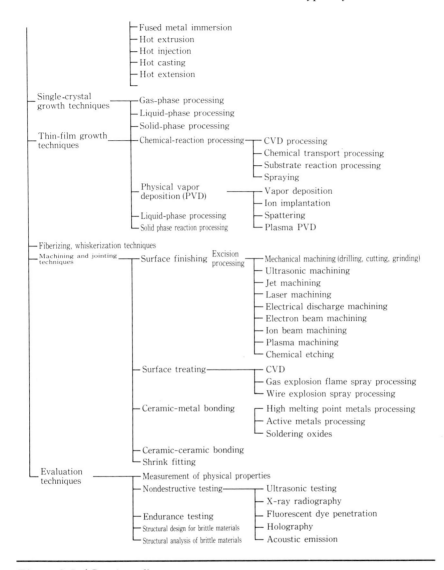

Figure 2.5. (*Continued*)

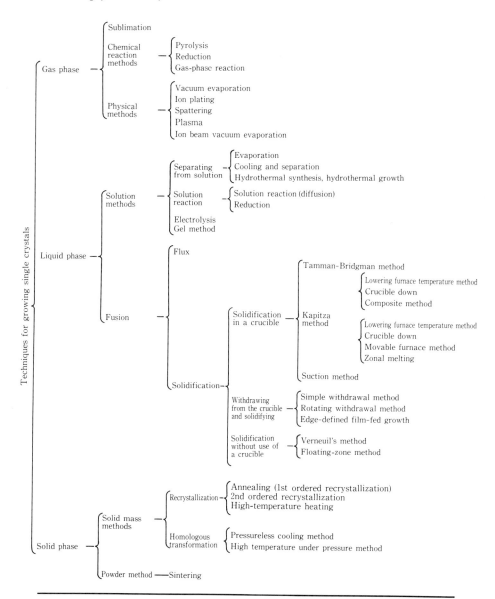

Figure 2.6. Single-crystal production technology in ceramics processing. Adapted from Taki (1976, p. 35).

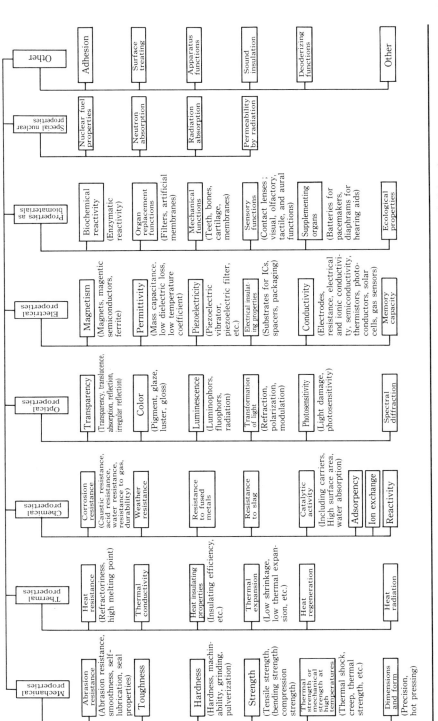

Figure 2.7. Classification of ceramics by their properties. Revised from materials provided by the Diamondosha keiei kaihatsu jōhō henshubu.

Table 2.3 Properties and Uses of Technical Ceramics[a]

Field	Properties	Uses
		Optics, electronics, magnetism
Electroceramics	Good insulating properties	Integrated circuit packages, integrated circuit substrates, heat-radiating insulating substrates
	Ferroelectric properties, permittivity	Image memory components, electrooptical polarizing components, high-volume capacitors
	Piezoelectric properties	Radiators, ignition components, radio wave filters, piezoelectric transisters, ultrasonic devices, electronic lighters, elastic surface wave components, electronic clocks
	Pyroelectric properties	Infrared detection components, thermography, detectors, special weapons
	Electronic radiation properties	Hot cathodes for the electron gun in television tubes, thermionic devices, electron microscopes, electron-beam welding, heat direct power generators, electron beam lithography equipment for VLSIs
	Semiconductor sensor properties	Resistance heating elements (high-temperature electronic furnaces), humidity sensors, thermistors (temperature control devices), pressure sensors, voltage-dependent resistors (varistors), self-regulating resistance heating elements (in electronic rice cookers, bedding dryers, or hair dryers), gas sensors (gas leak detectors)
	Ionic conduction properties	Oxygen sensors (air/fuel ratio control devices in automobile engines), blast furnace controls, sodium-sulfur batteries (for electric power equalization)
Optoelectroceramics	Fluorescence	Fluophors, materials used in color television tubes
	Polarization	Electrooptical polarization components
	Many questions remain for development in the 1980s.	The 1980s offer great promise for such developments as a photoelectric conversion component.
Optoceramics	Transparency	Transparency with heat resistance and corrosion resistance (high-voltage sodium lamps), spy holes for kilns, windows for nuclear reactors, transparency to visible light (nonfogging glass)
	Reflection of light	Heat resistance and metallic characteristics
	Reflection of infrared light	Transparance to visible light but reflecting infrared (energy-saving window glass)

Magnetic ceramics	Transmission of light	Optical fibers for telecommunications, optical communication cables, gastroscopes, optical energy transmission fibers
	Soft magnetism hard magnetism	Memory components for computers, magnetic cores for transformers, magnetic tape, magnetic disks, rubber magnetics, stereo pickups, magnetic heads, magnetic cash cards, magnetic door seals for refrigerators
Thermal properties	Thermal conductivity	Insulating (radiating) substrates for integrated circuits
	Thermal insulating property	Heat-resistant insulators, lightweight insulators, fireproof wall materials, energy-saving furnace materials
	Heat resistance	Heat-resistant structural materials, high-temperature furnaces, fusion reactor materials, nuclear reactor materials
Mechanical properties Engineering ceramics	High strength, resistance to abrasion, no expansion and contraction	Ultraprecision all-ceramic lathes and machine tools, measuring instruments, and wire drawing dies
	High strength, heat resistance	High performance, highly efficient automobile engines, gas turbine vanes, diesel engines, Stirling engines, heat-resistant tiles for the space shuttle
	High strength-to-weight ratio	Automobile parts, manmade satellite parts, rocket fuselages, airplane fuselages
	Great elasticity	Golf clubs and shafts, tennis rackets, pole-vault poles, fishing poles, various springy materials
	Ultrahardness	Grinding materials, cutting tools, abrasives, bits for excavating, scissors, knives
	Lubricating properties	Bearing materials, high-temperature lubricants
		Biological and chemical
Bioceramics	Bone compatibility (replacement for bone material)	Artificial bones, artificial teeth, artificial joints (surgical knives)
	Carrier properties	Carriers for immobilized enzymes, carriers for catalytic agents, control devices for biochemical reactions, linings for combustion chambers
	Corrosion resistance	Physics and chemistry apparatus, chemical engineering apparatus, nuclear power-related materials, linings for chemical apparatus
	Catalytic properties	Catalysts in water-gas reactions, heat-resistant catalysts, and catalysts in C_1 chemical reactions

[a] Adapted from Ho (1982, p. 79).

3
Synthetic Raw Materials for Ceramics

Akira Nagai and Yoshitaka Kimura
Showa Denko, K. K.
Minato-ku, Tokyo 105, Japan

I. INTRODUCTION

In ceramics, unlike metallic materials or synthetic resins, the character and lineage of the metallic oxides and nonoxides that make up the powder of the initial material have, through the processing method used, a decisive influence on the properties of the end product. *Character* and *lineage* refer to the composition of the materials, the impurities they contain, and the diameter, distribution, form, and state of cohesion of the particles of the powder form in which the material is used. *Processing* includes the forming method, the sintering aid used, the sintering temperature, the duration of sintering, and the sintering atmosphere.

Each particle of the sintered body forms grain boundaries. In ceramics, one special characteristic can be described by the integration of interaction between the bulk particles and the particle boundaries. Defects, of a microscopic order, of pores left within the particles and on the particle

Note: Akira Nagai wrote sections I, II, and III,A. Section III, B is by Yoshitaka Kimura.

Table 3.1 Ceramic Characteristics

Character of the powdered material	Character of the green body	Character of the sintered body	Characteristics required
Chemical composition of the material	Density of the green body	Composition and purity	Heat-related qualities
Purity	Uniformity	Crystalline form	Heat resistance
Particle structure and activity	Anisotropy	Lattice defects	Thermal shock resistance
Crystalline form	Residual stress	Micro structure	Low thermal expansion
Specific surface	Dry strength	Single phase	Thermal conductivity
Surface energy	Forming auxilliary	Density (porosity)	Strength at high temperatures
Ion-exchange capability	Sintering auxilliary	Particle-size distribution	Thermal insulating ability
Interior strain		Particle shape	Mechanical qualities
Point defects		Pore distribution (within particles, outside particles)	Hardness and abrasion resistance
Surface defects			Toughness
Lack of uniform composition			Lubricity
Form of the primary particles		Pore shape and its distribution	Workability
Grain-size distribution of the primary particles		Ratio of open and closed pores	Electrical qualities
Form of the secondary particles			Electrical resistance
Grain-size distribution of the secondary particles		Grain-boundary segregation	Magnetic qualities
Adhesive power		Grain-boundary stress	Optical qualities
		Grain anisotropy	Transparency
		Cracks and voids	Chemical qualities
		Multiple phase	Resistance to corrosion
		Phase types and sizes	Nuclear qualities
		Shapes and distributions	Other qualities
			Anisotropy
			Dimensional precision
			Surface smoothness

boundaries and of deposition and segregation of impurities are characteristic flaws in ceramics and major causes of reduction in reliability. The relationship between these factors and the quality of the resulting ceramics is becoming increasingly clear. Moreover, the example of alumina, the archetypical ceramic material for which extreme micro control of the particles making up the raw material is necessary, shows a record of progress in producing higher quality alumina ceramics. This progress was achieved through constant improvements in the alumina powder production process that created a better raw material.

As the history of alumina suggests, selecting and using good raw material powders is a critical first step in producing good ceramics. This idea is now widely accepted, but ceramists must remember that a good raw material powder cannot be evaluated in isolation. Rather, one should take into account the relationships of the various factors, and their interactions, in all the processes used in creating ceramics, including the molding and firing in deciding among the appropriate choices (Yōgyō Kyōkai, 1979, p. 181). Table 3.1 illustrates some of these complex relationships.

II. GENERAL ISSUES CONCERNING POWDERS FOR CERAMICS

A. EFFECTS OF IMPURITIES

The principal chemical qualities in ceramic raw materials are the chemical species of the impurities, the amounts present, and the state in which they exist. The influence of impurities is not a simple function of the amount in the ceramic material. To gauge their effect, it is necessary to investigate the source of the raw material minerals and the production process, particularly the heating history. The way impurities are distributed, whether they are sited on the surface or on the particle boundaries, greatly affects the sintering action.

For instance, in alumina ceramics, which are used in electronic devices, the presence of alkaline metallic ions is to be avoided because they lower the ceramic product's electrical insulating ability (Figure 3.1). Much research on removing alkaline metallic ions while preparing the main raw material, alumina, has led to a recent upgrade in the quality of Bayer-process alumina, with Na_2O reduced the 100 ppm. Fukuura and Asano (1983, p. 116) have reported the effect of Na_2O content on volume specific resistance. In addition, alpha rays emitted by impurities (U,Th) within the ceramic package may induce soft errors in semiconductor memory, a problem reported by May et al (1978, p. 33). The source of the problem is thought to be extremely small amounts of uranium (on the order of

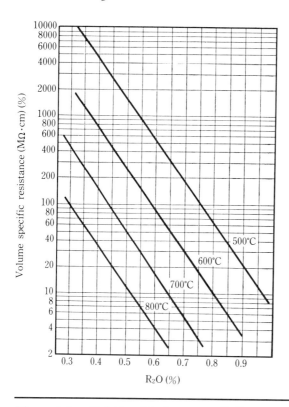

Figure 3.1. Relationship between alkaline content and insulation resistance in 90% alumina ceramics. From Fukuura and Asano (1983).

1 ppm). Research on eliminating that uranium has continued, leading to the development of alumina with an alpha ray count of less than 0.01 $c/cm^2 \cdot h$. A similar issue has arisen recently concening SiC and Si_3N_4. The bonding state of C and the O_2 analysis are said to be problematic.

B. CHARACTERISTICS OF POWDERS

When specifying the basic qualities that a raw material for a ceramic sintering body must possess, authorities frequently state that a powder must be as fine as possible, with equal particle diameters, and isodiametric. Both the material's microscopic characteristics as particles—individual particle size and distribution, the size of the aggregate particles, their distribution, and their strength—and its macroscopic nature as a particulate substance—bulk density, compressibility, fluidity, volume of absorbed liquid—are important data. The properties required in the finished ceramics regulate the characteristics sought. These characteristics must be strictly controlled during the manufacture of these powders.

III. MAJOR MATERIALS FOR CERAMICS

The following discussion covers synthetic materials for ceramics which are likely to become even more important in the near future: alumina, zirconia, magnesia, silicon carbide, and silicon nitride. Table 3.2, 3.3, and 3.4 present a comparison of the characteristics of these major ceramic materials.

A. OXIDES

1. Alumina (Al_2O_3)

Alumina is the most widely used of the synthetic raw materials for ceramics. The production technology for this material and research and development in the alumina ceramics field have accumulated and borne-fruit over many years. In addition, the cost of the material and its reliable supply give alumina, in its overall value as a material, a leading position which it is unlikely to lose soon. In the near future, improvements in the characteristics of alumina as a material will permit prediction of new properties and fields of application to be developed.

Many of the oxide ceramics have strong ionic bonds. Of these, Al_2O_3 has the most stable physical properties, with excellent heat and corrosion resistance. A loss of strength at high temperatures is unavoidable, but a variety of applications are being developed for it as machine parts used in a lower temperature range, such as cutting tools, pumps, and valves.

Although alumina can be represented simply by the chemical formula Al_2O_3, its nature varies considerably depending upon, for instance, its crystalline form, the impurities present, and the particle diameter. Since required physical properties also vary according to the intended use, many different types of alumina powders are commercially available. The following discussion of the method of manufacture for alumina, its types, uses, and physical properties is intended for reference in choosing the specific alumina that matches the intended use from among all those commercially available.

a. THE ALUMINA MANUFACTURING PROCESS. The production of alumina for ceramics is based on the Bayer process, which was developed

Table 3.2. Comparison of Basic Characteristics of Ceramic Materials: Physical and Mechanical Characteristics

| Material | Sp. gr. | Hardness | | | | Strength (kg/cm²) | | | | |
| | | Mohs (new) | Knoop (kg/mm²) | Vickers (kg/mm²) | Rockwell A | Bending | | Compression RT (1,000°C) | Sheer 1000°C | Young's modulus (kg/cm² × 10⁶) |
						RT	1000°C			
Alumina (Al₂O₃)	3.98	12	2,100–2,500	2,300–2,700	95	3,000–4,000 (6,000)	—	2,800–3,500	1,500	3.5–4
Beryllia (BeO)	3.02	9– (12)	1,200	—	—	1,500–2,000	—	—	1,500	4 3
Cerium oxide (CeO₂)	7.13	(6)	—	—	—	—	—	—	—	2
Chromia (Cr₂O₃)	5.21	(12)	—	—	—	—	—	—	—	—
Magnesia (MgO)	3.58	6	600–900	—	—	1,600–2,800	—	5,000–6,000	1,000	2–3
Silica (cristobalite) (SiO₂)	2.32	6.5	—	—	—	—	—	—	—	—
Quartz (SiO₂)	2.65	7	800–950	1,000	—	—	—	20,000	—	1
Amorphous silica (silica glass) (SiO₂)	2.20	7	—	500–700	—	500–1,000	—	7,000–19,000	1,000	0.7
Titania (rutile) (TiO₂)	4.24	7–9	(single crystal) 1,000	—	—	700–1,700	—	2,800–8,400	1,200	1–2

Material									
Stabilized (cubic) zirconia (ZrO_2)	6.27	8–9	—	—	—	1,800–8,000	—	10,000–30,000	1,400	1.5–2
Monoclinic zirconia (ZrO_2)	5.56	8–9	—	—	—	1,800–8,000	—	10,000–30,000	—	2.5
High-strength zirconia (ZrO_2)	5.7–6.1	—	—	1,300–1,500	91	10,000–15,000	—	—	—	2
Mullite ($3Al_2O_3 \cdot 2SiO_2$)	3.16	8	700–1,400	—	—	1,100–1,900	980	4,000–6,000	850	0.5–1.5
Spinel ($MgO \cdot Al_2O_3$)	3.58	7	—	1,540	—	1,500–1,7000	—	17,000	—	2.6
Cordierite ($2MgO \cdot 2Al_2O_3 \cdot 5SiO_2$)	2.0–2.5	7	—	—	—	1,200	1,000–1,2000	3,500–6,800	350	1.5
Silicon carbide (α-SiC)	3.22	13	2,500–3,000	2,000–3,000	—	4,500–8,000	—	6,000–42,000	—	4–6
Titanium carbide (TiC)	4.94	9–12	2,500	3,200	93	(25% porosity)	—	7,600–13,800 (8,800)	1,400	3–4
Boron carbide (B_4C)	2.52	—	2,800	—	—	3,400 / 2,100	2,100	29,000–18,000	—	—
Tungsten carbide (WC)	15.6	>12	1,900	1,800–2,200	81	3,500–8,500	(630)	27,000–36,000 (14,000)	—	7
Graphite for electrodes (C)	2.23	1–2	—	—	—	80–300	—	300–650	—	0.05–0.12
Hexagonal boron nitride (h BN)	2.27	2	—	—	—	500–800	—	700–1,000	—	0.5–0.8
Silicon nitride (Si_3N_4)	3.17	>12	2,400–2,800	1,700–2,700	93	5,000–10,000	—	5,000–8,000	—	3–4

Table 3.3. Comparison of Basic Characteristics of Ceramic Materials: Thermal Characteristics

Material	Melting point (°C)	Sp. heat (cal/g · °C)	Thermal conductivity (cal/cm · sec · °C)				Coefficient of thermal expansion (× 10⁻⁶/°C)		
			RT	100°C	400°C	1000°C	RT	400°C	1000°C
Alumina (Al_2O_3)	2050	0.25	single crystal 0.095	100% density 0.07	0.03	0.015	6–9	7	9
Beryllia (BeO)	2550	0.24	—	0.55	0.22	0.05	6–9	8	9
Cerium oxide (CeO_2)	>2660–2800	0.10	—	86–92% density (0.03)	—	—	12	—	—
Chromia (Cr_2O_3)	1990–2260	0.17	—	—	—	—	5.5–9	—	—
Magnesia (MgO)	2800	0.20–0.29	single crystal 0.17	100% density 0.14–0.08	0.04	0.017	11–15	13	15
Silica (cristobalite) (SiO_2)	1720	0.2	—	0.003–0.03	—	—	5	—	—
Quartz (SiO_2)	1610	0.2 //	single crystal (//C axis) 0.017	—	—	—	17–30	—	—
Amorphous silica (silica glass) (SiO_2)	—	0.2	—	0.002–0.004	0.004	—	0.5–1.4	—	—
Titania (TiO_2)	1840	0.17–0.21	—	0.008–0.015	0.007–0.009	0.008	7–9	—	—
Stabilized (cubic) zirconia (ZrO_2)	2715	0.12–0.17	—	0.005	0.005	0.005	7–10	—	—

Material									
Monoclinic zirconia (ZrO_2)	—	—	—	0.005	—	—	—	—	—
High-strength zirconia (ZrO_2)	—	—	—	0.004	—	—	8–9	—	—
Mullite ($3Al_2O_3 \cdot 2SiO_2$)	1830	0.2	—	0.007–0.015	0.011	0.010	4.5–5.5	—	—
Spinel ($MgO \cdot Al_2O_3$)	2135	0.20	—	0.04	—	—	8–9	—	—
Cordierite ($2MgO \cdot 2Al_2O_3 \cdot 5SiO_2$)	1460	—	0.005–0.02	—	—	—	—	RT–1000°C 1.4–2.1	—
Silicon carbide (α-SiC)	decomposition 2220	0.16	0.16–0.4	—	—	0.08–0.3	3.5–5.5	—	—
Titanium carbide (TiC)	3400–3500	—	—	0.04–0.06	—	—	7.4	—	—
Boron carbide (B_4C)	2450	—	—	0.06–0.11	—	—	4.5	—	—
Tungsten carbide (WC)	decomposition 2600	—	—	0.07	—	—	4–6.2	—	—
Graphite for electrodes (C)	sublimation 3650	—	—	0.25–0.58	0.27	0.15	1–3.5	—	—
Hexagonal boron nitride (h BN)	decomposition 3000	—	—	0.04–0.1	—	—	1.5–3	—	—
Silicon nitride (Si_3N_4)	1900	0.15–0.2	—	0.04–0.06	—	—	2–3	—	—

Table 3.4. Comparison of Basic Characteristics of Ceramic Materials: Electrical Characteristics

Material	Volume-specific resistance ($\Omega \cdot$cm)	Dielectric strength (kV/mm)	Inductance	tan δ (1 MHz, RT)
Alumina (Al_2O_3)	10^{14}–10^{16}	>20	5–10	4×10^{-4}
Beryllia (BeO)	>10^{14}	>10	4–6	6–38×10^{-4}
Cerium oxide (CeO_2)	—	—	—	—
Chromia (Cr_2O_3)	—	—	—	—
Magnesia (MgO)	>10^{14}	30	7–8	1–8×16^{-4}
Silica (Cristobalite) (SiO_2)	—	—	—	—
Quartz (SiO_2)	—	—	—	—
Amorphous Silica (Silica glass) (SiO_2)	10^9	36–500	3.5	—
Titania (TiO_2)	>10^{13}	10–20	15–80	2–500×10^{-4}
Stabilized (cubic) zirconia (ZrO_2)	MgO, Y_2O_3 Solid solution $\rightarrow$ high conductivity	—	10 GHz 17	—
Monoclinic zirconia (ZrO_2)	>10^{13}	—	—	—
High-strength zirconia (ZrO_2)	>10^{10}	—	—	—
Mullite ($3Al_2O_3 \cdot 2SiO_2$)	>10^{14}	10–16	7	60–700×10^{-4}
Spinel (MgO $\cdot$ Al_2O_3)	>10^{14}	10	8	10^{-3}–10^{-4}
Cordierite (2MgO $\cdot$ $2Al_2O_3$ $\cdot$ $5SiO_2$)	>10^4	14	—	—
Silicon carbide (α-SiC)	100–200	—	—	—
Titanium carbide (TiC)	10^{-4}	—	—	—
Boron carbide (B_4C)	0.3–0.8	—	—	—
Tungsten carbide (WC)	5×10^{-5}	—	—	—
Graphite for electrodes (C)	0.4–1.2×10^{-3}	—	—	—
Hexagonal boron nitride (hBN)	>10^{14}	30–40	4–5	1–2×10^{-4}
Silicon nitride (Si_3N_4)	>10^{14}	—	9.4 GHz 7.5	—

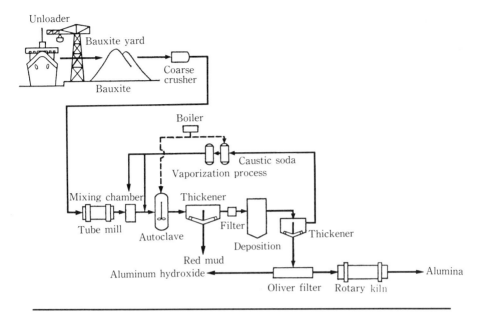

Figure 3.2. Flowchart of the Bayer process.

for refining aluminum (Figure 3.2). Alumina is produced using a partial modification of the basic Bayer process and separate firing and processing equipment. The basic raw material used is bauxite. After heat processing in caustic soda, the alumina is extracted. Then, following separation of the Fe_2O_3, TiO_2, SiO_2, and other insoluble residues, seed crystals are added and the ore is cooled and stirred, which produces aluminum hydroxide by hydrolytic deposition. Since K. J. Bayer discovered this method in 1888, there have been no essential changes in the method used to refine aluminum. Alumina is produced when this aluminum hydroxide is baked in, for instance, a rotary kiln. Alumina has an average particle diameter of 40 to 100 μm. Normally, α-Al_2O_3 is the principal crystal form. The alumina produced in this way contains approximately 0.3% Na_2O and 0.01% SiO_2 as impurities; Al_2O_3 with a purity of 99.6% or better is obtained. For many applications as a raw material for ceramics, the alumina can be used quantitatively at this level of purity.

However, for applications requiring a high degree of electrical insulating capability, such as electronic materials and sparkplugs, low-soda alumina, with less than 0.1% Na_2O impurities is used. Recently, 3N alumina (purity of 99.9% or better) has been developed by an improvement in the Bayer process; this alumina is used in high-strength ceramics such as ceramic tools. Production methods other than the Bayer process have developed alumina with the ultrahigh purity of 4N or 5N and with a fine crystalline form, the crystals 0.2 to 1.0 μm in diameter. These new products have been attracting interest as materials for alumina single-crystal or transparent ceramics. The new methods include

1. Pyrolysis of ammonium alum
2. Hydrolysis of organic aluminum compounds
3. Pyrolysis of ammonium dawsonite
4. Electric discharge of aluminum in water

b. TYPES OF ALUMINA CERAMICS AND THEIR CHARACTERISTICS. Figure 3.3 presents the production factors that determine the character of alumina. First, the impurities included in the aluminum hydroxide, the basic ingredient in alumina, and its particle structure influence the character of the alumina. They determine the behavior of the alumina during heating and its physical properties. Therefore, control of impurities and of particle structure is critical in the Bayer process. In addition, factors such as pyrolysis and firing conditions are strictly controlled. The primary factors in the nature of the resulting ceramics, the particle diameter, grain-size distribution, and form of the primary and secondary particles of α-Al_2O_3 and of the impurities, are decided. Finally, pulverizing and particle rectification processes are performed as needed.

Figure 3.4 graphs the relationship between the impurity level (Na_2O, by percentage) and the average diameter of the primary particles for several commercially available aluminas. These two characteristic values were chosen because elimination of these impurities that influence particle growth during sintering is definitely desirable and because the diameter

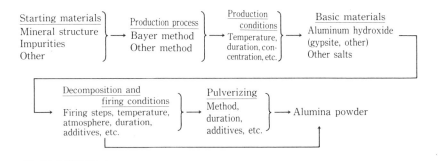

Figure 3.3. Production factors that determine the character of alumina.

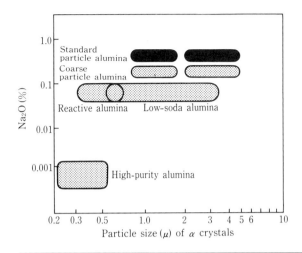

Figure 3.4. Relationship between particle size of aluminas and Na_2O.

size and uniformity of the primary particles of the alumina are major assessment factors in the characteristics of the end product.

In alumina powder, formed when pyrolysis of aluminum salts or hydroxides produces α-Al_2O_3, generally the relatively fine primary particles form secondary particles by bonding together in large numbers (See Figures 3.5, 3.6, 3.7, and 3.8). The fact that the finer the particle diameter of the ceramic raw material, the better it sinters is not confined to alumina ceramics. The point also follows from the theories of volume diffusion and grain-boundary diffusion (Kingery et al., 1976). In this case, in the pulverizing process normally adopted, the ease with which the secondary particles can be smashed and their capacity for recohesion are highly important characteristics. In normal molding powders, these recombined secondary particles are hard to break apart, and pores tend to remain between them when the object is sintered. It is difficult to remove these pores by sintering. Therefore, to achieve a sintered body with high density, the most appropriate values for the diameter of the primary particles must obtain. In the case of alumina, that value is assumed to be between 0.2 and 0.5 μm.

For pulverized alumina, the manufacturers perform pulverization processes appropriate for each application. The degree of pulverization can be selected almost at will down to the diameter of the primary particles. The physical properties and uses of commercially available alumina are given in Tables 3.5 and 3.6.

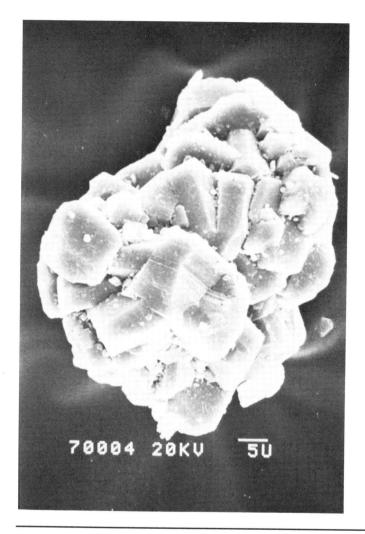

Figure 3.5. Standard particle of aluminum hydroxide.

2. Zirconia (ZrO_2)

Zirconia-based ceramics have also shown a dazzling development recently, in a variety of unique materials derived from the singular character of that material. The zirconia powder which first appeared was 95 to 98% ZrO_2 and was used principally in refractories. The main production method for this power is the dry method. Zircon sand, $ZrSiO_4$, is melted in an electric arc furnace, dispersing the SiO_2 to produce zirconia. For stabilized zirco-

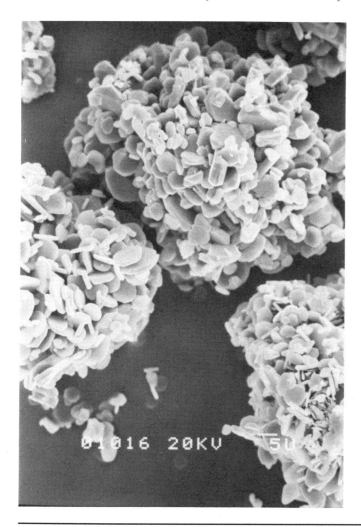

Figure 3.6. *Flowery-type standard alumina.*

nia, CaO is added during electromelting to stabilize the zirconia and to make the SiO_2 soluble.

Hydroxides, carbonates, and other precipitates are made from an acid solution of the products of the decomposition of zircon by soda. Light monoclinic zirconia is obtained by calcination of these precipitants; monoclinic heavy zirconia is obtained by refining the natural ZrO_2 ore, baddeleyite, after firing. Both are monoclinic systems which shift to a tetragonal system near 1100°C. Near this transition temperature, a large change in volume occurs, in which, therefore, CaO, MgO, or Y_2O_3 are

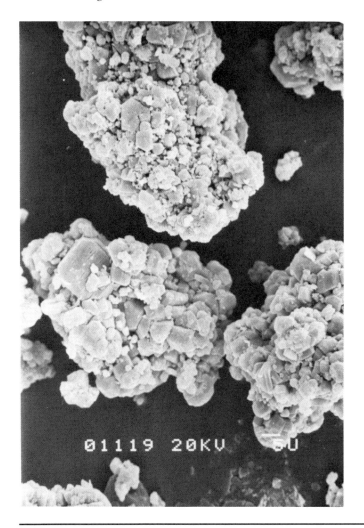

Figure 3.7. Sandy-type standard alumina.

dissolved. The resulting material, an isometric system with no abnormal expansion and contraction that will not change into a tetragonal system, is known as stabilized zirconia.

In a dimorphous composition, with both monoclinic and tetragonal crystals, it is possible to increase the ceramic material's strength and its thermal shock resistance by using the stress generated within the structure of the ceramic object. Such a form of zirconia, containing dimorphous crystals, which is not completely stabilized, is called partially stabilized

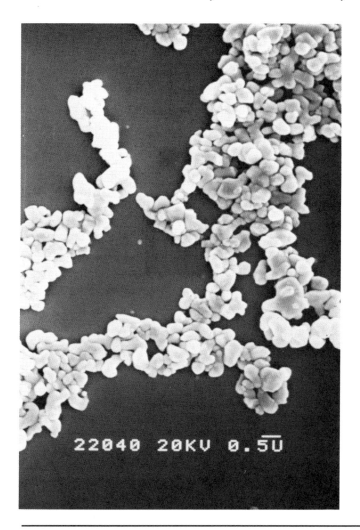

Figure 3.8. *Ultrahigh-purity alumina, BET 5 m^2/g.*

zirconia. It is highly suitable as a ceramic material (Scott, 1975, p. 1527). With Y_2O_3-PSZ (Y_2O_3-partially stabilized zirconia), the Y_2O_3 content has been reduced from the 8 mol% hitherto necessary to stabilize zirconia to the 3 to 4 mol% level. A remarkable increase in strength occurs with a composition of 3 mol% Y_2O_3, as shown in Figure 3.9 et al., 1981, p. 2).

Use of this powder makes it possible to obtain sintered bodies with a uniformly fine grain texture and close to the theoretical density without special processing of the raw material.

Table 3.5. Characteristics of Commercially Available Aluminas[a]

	Standard alumina		Coarse-grained alumina		Pulverized alumina		Low-soda alumina		Reactive alumina			High-purity alumina
Product name												
Showa Light Metal	A-12	A-14	A-12C	A-14C	A-42	A-43	AL-15	AL-13	AL-150SG	AL-160SG	AL-170	UA-5000 series
Sumitomo Aluminum	A-21	A-26	AC-21		AM-21	AM-27	AL-43	AL-41	AES-21	AES-11E	AES-22S	AKP-20[c] AKP-30[c]
Nippon Light Metal Mitsui Aluminum	A-11 MA-11	A-13 MA-20	A-21	A-23 MAS-20	A-31 MM-21 ground	A-32 MM-20 ground	LS-20[b]	LS-30[b]		XAL-23V[b]		
Alcoa	A-5		A-2	A-3	A-2	A-1, A-3	A-14		A-15SG	A-16SG	A-17	
Reynolds			RC-20	RC-23	RC-20 GF	RC-23 BM	RC-122	RC-152	RC-122 DBM	RC-172		
Chemical composition (%)												
Al_2O_3	99.6	99.6	99.7	99.7	99.6	99.6	99.8	99.9	99.9	99.9	99.8	Na 10 ppm
SiO_2	0.02	0.02	0.02	0.02	0.02	0.02	0.07	0.02	0.03	0.03	0.05	K 5
Fe_2O_3	0.01	0.01	0.02	0.02	0.01	0.01	0.01	0.02	0.02	0.02	0.02	Ca 2
Na_2O	0.35	0.35	0.25	0.25	0.35	0.35	0.05	0.07	0.05	0.05	0.06	Mg 1
L.O.I.	0.10	0.35	0.10	0.40	0.15	0.4	0.10	0.10	0.5	0.7	0.2	Fe 4
Grain size												Si 6
Average particle size (μm)	60	60	75	80	4.0	1.8	60	60	2.0	0.6	2.0	Ga 2
+74 μm(%)	30	30	50	55								Cr 2
+44 μm(%)	75	75	85	90	0.1	4						Al_2O_3 99.997%
crystal size (μm)	5	5	5		4	1.5	2.0	2.0	2.0	0.4	2.5	0.03 ≈ 0.5
Specific gravity	3.96	3.94	3.96	3.94	3.95	3.94	3.97	3.96	3.92	3.91	3.93	3.42 ≈ 3.97
Bulk density (loosed)	0.7	0.9	0.8	0.9	0.8	0.7	0.9	0.9	1.0	1.0	1.1	0.2 ≈ 0.6
Bulk density (tapped)	1.0	1.1	1.1	1.1	1.4	1.2	1.2	1.2	1.9	1.4	1.8	0.4 ≈ 1.0
Bulk density (pressed) (1t/cm²)					2.25	2.05			2.60	2.25	2.55	1.25 ≈ 1.9

[a] Values of characteristics for Showa Light Metal products.
[b] Nippon Light Metal product.
[c] Sumitomo Chemical products.

Table 3.6. Uses of Alumina[a]

	Ceramics										Refractories						Abrasives			Other						
	Ceramics for physics and chemistry	Abrasive-resistant ceramics	Saggers	Electrical insulators	Spark plugs	Electronic components	Ceramic tools	Transparent, polycrystalline substances	Single-crystal substances	Bioceramics	Electrofused alumina	Sintered alumina	Synthetic spinel	Shaped refractories	Monolithic refractories	Ceramic fibers	Abrasives for hard materials	Abrasives for soft materials	Precision abrasives	Raw materials for glass	Welding rods	Flux	Placing sand, parting compound	Catalysts	Resin filling agents	Paints
Standard particles																										
A-12	○	○	○	○							◎		◎	○		○				◎		◎				
A-13		○	○	○							○					○				○		◎	○			
A-14													○													
Coarse particles																										
A-12C			◎								◎	◎	○			◎					◎		○			
A-14C											○	◎				◎										
Fine particles																										
A-42 series	○	○	○	◎									◎	◎	○							○	○	○	○	◎
A-420		○												○	◎		○							○		
Ultrafine particles																										
A-43	○		◎			○													○							
A-50 series						○											◎	◎								
Low soda																								○		

[a] Double circles, primary use; single circles, frequent use.

(*continued*)

Table 3.6. (*Continued*)

	Ceramics										Refractories						Abrasives			Other						
	Ceramics for physics and chemistry	Abrasive-resistant ceramics	Saggers	Electrical insulators	Spark plugs	Electronic components	Ceramic tools	Transparent, polycrystalline substances	Single-crystal substances	Bioceramics	Electrofused alumina	Sintered alumina	Synthetic spinel	Shaped refractories	Monolithic refractories	Ceramic fibers	Abrasives for hard materials	Abrasives for soft materials	Precision abrasives	Raw materials for glass	Welding rods	Flux	Placing sand, parting compound	Catalysts	Resin filling agents	Paints
AL-13	◎	◎				○																				
AL-13PC	○			○	◎	◎																			○	
AL-15 series	○	○			○	○																				
AL-30	○					◎																			◎	
AL-43						◎																	◎		◎	
AL-45 series																										
Reactive	○						○			○																
AL-150SG		◎					○			◎				○		○		○							◎	○
AL-160SG						○	○			○				○	◎										◎	
AL-170															◎											
A-171															◎									○	○	
A-172			◎																							
High-purity																			◎							
UA series						◎	◎	◎	◎	◎									○					○		

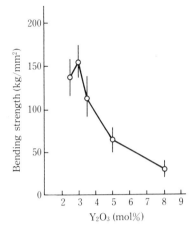

Figure 3.9. Strength of Y_2O_3-PSZ hot-pressed at 1400°C for 1 hour, n = 30. From Masaki et al. (1981).

3. Magnesia (MgO)

Magnesia has a high melting point, 2800°C, and is an abundant natural resource. It reacts, however, with water, carbon dioxide, and acids, turning into magnesium hydroxide, magnesium carbonate, and other magnesium salts. Magnesia, unlike alumina or silica, does not have very stable chemical properties. If magnesium hydroxide, which is the raw ingredient for MgO, is heated gradually, fine crystals grow within the magnesium hydroxide. The specific surface of magnesium oxide crystals is large, several hundred square millimeters per gram, and it is strongly reactive; MgO is obtained by its decomposition. If the MgO is heated further, its activity is reduced rapidly but even at 1200°C considerable activity remains. This highly active MgO is called active magnesia or light magnesia. For a better sintered body, pressure forming in this state combined with high-temperature sintering are the methods chosen, as in sintering other materials.

Magnesia clinker, used as a refractory material, is sold commercially with a density at 96% of theoretical. After products molded of magnesia clinker or light magnesia are melted in an electric arc furnace, recrystallized, electromelted magnesia is formed.

Table 3.2, 3.3, and 3.4 summarize the general properties of MgO. Its particularly striking characteristics, however, are its high melting point, its rather large coefficient of expansion under heat, and its high electrical resistance.

Table 3.7 presents examples of the qualities in which magnesia clinker is available and Table 3.8 does the same for electromelted magnesia (Takamiya, 1982, p. 823).

The ceramic characteristics of MgO vary considerably, depending upon the type of mother salt and the calcination temperature. Therefore, these

Table 3.7. Sample Qualities of Magnesia Clinker[a]

Composition (%)	A	B	C	D	E	F
MgO	99.2	98.5	98.0	97.8	95.0	72.3
CaO	0.46	0.90	1.30	0.90	1.0	25.4
SiO_2	0.18	0.30	0.40	0.90	2.7	0.64
Fe_2O_3	0.04	0.07	0.08	0.07	0.3	0.94
Al_2O_3	0.04	0.07	0.08	0.07	0.6	0.26
B_2O_3	0.03	0.07	0.05	0.10	0.2	0.18
Bulk density	3.30	3.40	3.45	3.30	3.20	3.35

[a] From Takamiya (1982, p. 823).

Table 3.8. Sample Qualities of Electrofused Magnesia[a]

	Single crystals	Uses in electrical heating elements	Refractory uses
MgO (%)	≥99.9	≥99.2	≥97.2
CaO (%)	≤0.006	≤0.40	≤1.5
SiO_2 (%)	≤0.002	≤0.10	0.50
Fe_2O_3 (%)	≤0.008	0.09	0.35
Al_2O_3 (%)	≤0.003	≤0.20	0.30
B (ppm)	≥5	125	900

[a] From Takamiya (1982, p. 823).

factors make selection of the raw material a critical point (Hamano, 1983, pp. 11–16).

B. THE NONOXIDES

1. Silicon Carbide (SiC)

Silicon carbide (SiC), an artificial material with a strong covalent bonding structure, was accidentally synthesized by E. G. Acheson in 1891. It is harder and more heat resistant than Al_2O_3, and it also demonstrates outstanding resistance to corrosion. Therefore, it is widely used as both an abrasive and a refractory material. It is also used as a metallurgical additive in iron- and steelmaking, and, putting to use its electrical properties, in heating elements and in parts for electrical circuits (varistors and arrestors). In Japan, accompanying the development of the steel industry, an annual production of more than 70,000 metric tons of SiC has been achieved. The production method uses an electric furnace, basically following the process Acheson discovered. That is the only method of mass production.

In recent years, SiC has attracted attention as a new ceramics material. Its development as a ceramic material is proceeding, with the expectation that SiC will manifest even more advanced features. For these purposes, production processes other than the Acheson method are also employed.

a. TYPES AND CRYSTAL FORMS. Silicon carbide has two crystal forms: α-SiC, which belongs to either a hexagonal or a rhombohedral phase, and β-SiC, which belongs to a cubic phase. Polytypes of silicon carbide are represented by the way in which the layers formed by SiC tetrahedrons are stacked one upon another and by the number of layers in a repeating unit of SiC layer structure. In SiC crystals synthesized in the Acheson furnace, polytypes such as $4h$, $6h$, $15r$, or $3c$ are commonly observed (Hino, Iwama, Takaya, and Ichikawa, 1978, p. 115; Kawamura and Iwama, 1977/1978, p. 33). Their distribution varies according to the quality of the raw material used and the production conditions. The h, r, and c refer to hexagonal, rhombohedral, and cubic, respectively, and the figures indicate the number of layers in a unit cell. All are polytypes of α-SiC, except for the $3c$, which belongs to β-SiC.

The α form is a high-temperature stable form which is generated in the high-temperature zone of 1800° to 2000°C and above when synthesized in a reaction furnace. Obtained on a commercial scale as comparatively largely developed crystalline particles with a high degree of hardness, α-SiC exhibits good thermal stability, high heat resistance, and outstanding corrosion resistance. It is utilized in the fields described earlier. The β form is a low-temperature form generated in the low-temperature zone of 1500° to 1600°C. It can be obtained through a variety of methods as an extremely fine powder rich in activity. The β form has a low degree of hardness and, at 1800° to 2000°C, it begins an irreversible transition to the α form. It is also easily oxidized. Such characteristics have made it unsuited for the applications for which SiC, in the α form, has been used in the past. Industrial production has begun, however, using specialized production facilities, for the β form as a sintering raw material powder.

b. PRODUCTION OF SiC FINE POWDER FOR USE IN THE NEW CERAMICS. Recent progress in sintering technology has made it possible to achieve high-density sintered bodies of SiC, despite its low degree of sinterability. Such processes, however, require a high level of purity and a sufficiently fine powder—with diameters at the submicrometer level. The production methods used are either to micronize—render even finer—the α-SiC powders produced conventionally in the Acheson furnace or to synthesize fine particles of β-SiC directly.

(i) Alpha Silicon Carbide. As Figure 3.10 illustrates, the raw material, a mixture of silica and coke, is packed between fixed electrodes a certain

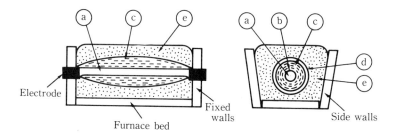

Figure 3.10. Sketch of an SiC (Acheson) production furnace. a, graphite resistance core; b, decomposition-formed graphite; c, α-SiC ingot; d, unreacted materials and β-SiC; e, unreacted materials.

distance apart in an electric resistance furnace (an Acheson furnace), with the electrodes connected by graphite powder. An electric current is passed between the poles, heating the furnace and causing the reaction to advance. The reaction within the furnace is complex but, in general, following the formula

$$SiO_2 + 3C \rightarrow SiC + 2CO$$

an ingot zone of α-SiC, passing through the β-SiC stage, is formed in the high-temperature zone around the graphite core. Since the further from the electrode core the lower the temperature of the zone, a cross section of the interior of the furnace after the reaction reveals an ingot zone of α-SiC with a band of β-SiC outside it; further outside there remains a layer of the unreacted raw-material mixture. The lumps of α-SiC are selected out and put through the processes of crushing, decarburizing by washing, removing iron, and classification, to arrive at the product with the desired particle size (Hino, Iwama, Takaya, and Ichikawa, 1978, p. 115; Kawamura and Iwama, 1977/1978, p. 33). To yield a fine powder with a high level of purity, a product with the appropriate grain size of green silicon carbide. (GC) is used. It is normally produced by using raw materials of high purity, volatilizing impurities out of them by adding salt and wood chips, and passing the resultant product through specialized fine grinding, classification, and refining processes.

Ball mills, vibrating mills, and jet mills are used for the fine grinding. Since the particle shape is an important characteristic of sintering materials, in choosing the type of mill and the conditions for grinding, care must be taken to consider, apart from the aspect of grinding efficiency, the shape of the ground material produced. Generally, the dry ball mill and the vibrating mill produce spherical particles with few corners. The wet ball mill and the jet mill produce angular particles. Even among commercially available products, considerable differences in particle shape are to be seen.

There are two classification methods used: pneumatic and wet. The pneumatic classification method includes a variety of types, but they are inferior to the wet method in classification accuracy, though superior in classification efficiency. It is difficult to avoid the admixture of coarser particles when using the pneumatic classification method. The wet classification method includes several types such as sedimentation classification by gravity, hydraulic classification (elutriation tube), and centrifugal classification. In most cases, the wet method has the advantages of higher classification accuracy and lower equipment costs. At the same time, however, it has the disadvantages of requiring a drying process and of a lower classification efficiency than in the pneumatic method.

(ii) *Beta Silicon Carbide.* The several methods for producing fine particles of β-SiC are discussed next.

1. Carboreduction of silica: Following the same reaction formula as in the Acheson method, β-SiC is synthesized by heating a mixture of fine particles of both silica and carbon in the inert atmosphere of a furnace. It is necessary, however, to control the reaction temperature to keep it in a lower zone (1500° to 1700°C) to avoid formation of the α form of SiC. Continuous production in a vertical-type furnace is used. The reaction product is obtained as aggregates of fine particles. Since, however, it contains silica and carbon as unreacted substances, a refining process is required to remove them (Enomoto, 1982, p. 828).

2. Silicification of carbon (direct reaction between silicon and carbon): Powdered silicon and a carbon source such as carbon black are made to react directly at a comparatively low temperature, as given in the following formula (Hase and Suzuki, 1978, pp. 541, 606):

$$Si + C \rightarrow SiC$$

3. Gas-phase reaction, gas-phase pyrolysis: Gases such as $SiCl_4$ or SiH_4 are made to react with hydrocarbons, or gases such as CH_3SiCl_3 or $(CH_3)_4Si$ are made to undergo pyrolysis, yielding a fine powder. The heat sources used include plasmas and lasers (Katō, Okabe, Hōjō, 1980, p. 32; Okabe, Hōjō, Katō, 1980, p. 188).

c. *SILICON CARBIDE POWDER CHARACTERISTICS.* The characteristics of fine SiC powders marketed commercially are presented in Table 3.9.

1. Chemical composition: Although the method of analysis itself is somewhat problematic in the submicrometer region, due, among other factors, to an increased oxygen content in the powder, 97 to 98% pure SiC

Table 3.9. Examples of Characteristics of Commercially Available SiC Fine Powders

	Company A product				Company B product		Company C product	Company D product		Company E product
	A-1	A-2	A-2S	B-1	A-10	B-10	Ultra-fine	UF 15	UF 10	#10,000
Crystal form	α	α	α	β	α	β	β	α	α	α
Chemical composition (%)										
SiC	97	98	96	96	97	95	95	96	97	95
Free C	0.8	0.5	0.6	1.0	1.0	1.9	1.4	0.4	0.5	0.8
SiO$_2$	0.7	0.3	1.0	0.2	0.8	0.1	0.2	0.5	0.4	1.9
Total Al	0.01	0.01	0.01	0.05	0.10	0.04	0.10	0.06	0.07	0.05
Total Fe	0.04	0.07	0.27	0.04	0.01	0.02	0.04	0.01	0.02	0.08
Particle-size distribution 2 μm	0	15	5	0	18	10	8	0	1	0
Cumulative wt (%)										
1	5	40	30	1	40	31	26	4	20	11
0.4	60	75	70	48	68	63	69	52	69	76
Average particle size (μm)	0.4	0.7	0.6	0.4	0.7	0.6	0.6	0.4	0.6	0.6
Specific surface area (m^2/g)	15	9	11	18	11	15	16	15	10	14
Pressed bulk density	1.9	2.0	1.9	1.9	2.0	2.0	1.8	1.6	1.7	1.6

is obtained. In this connection, the GC (green silicon carbide) used as abrasive grains has a purity of 99% and above. Of the impurities found in SiC, free carbon and free SiO_2 are common, along with small quantities of free silicon, aluminum, iron, and other constituents. Differences in the proportions of such impurities are recognized in different products. These impurities are thought to affect the sintering properties of the powder and the properties of the sintered body, but there is no technically established explanation of their effects.

Here the SiC analysis value is calculated from total carbon minus free carbon, as specified in Japanese Industrial Standard (JIS) R 6124. Since, during combustion in an oxygen flow, finely powdered SiC is also partially oxidized, the free carbon content must be adjusted to compensate for that amount. For metallic impurities, the measured values are obtained by the inductively coupled plasma method.

2. Particle-size distribution and specific surface area: The particle-size distribution is measured using the centrifugal sedimentation method (Disk Centrifuge MK III, Joyce-Loebl). All the products have an average particle size of $1\mu m$ or less, but the proportion of coarser particles larger than a micrometer varies. In the finest powders, only 1 to 5% of the particles have particle sizes of 1 μm or more, with the largest particles 2 μm or less in size. The specific surface area, measured with the BET method, is 15 m^2/g and above for the finer powders. In general for powdered materials, the finer the particle size and the larger the specific surface area, the higher the density of the sintered body that can be obtained. Conversely, the density of the green body is lower, so that shrinkage during sintering is greater. Fine powders with a specific surface area larger than necessary are, contrary to expectation, difficult to handle (Böcker, Landfermann, and Hausner, 1981, p. 37), and at present a specific surface area of about 15 m^2/g is recommended.

3. Particle shape: Examples of the particle shape of both α- and β-SiC fine particles are given in the scanning electron micrographs in Figure 3.11. The micrograph of β-SiC shows a preponderance of comparatively round particles, but the α-SiC particles are irregularly shaped, with many rather long, narrow, and angular forms, due to its being milled from coarser particles. As can be seen from the micrograph, however, depending upon the milling conditions, it is possible to produce a fine powder in a blocky shape with an elongation ratio of nearly 1. In this case, the specific surface area is lower for a given particle size.

4. Packing characteristics of the powder: The pressed density is measured by placing a 15-gram sample in a 30 mm ϕ die and press forming it by the application of 1 metric ton/cm^2 of pressure for 30 seconds. This value varies according to the mean particle size and the particle-size distribution. For a given particle-size distribution, the closer the particle shape is to spherical, the higher this value (note also the specific surface area). It can

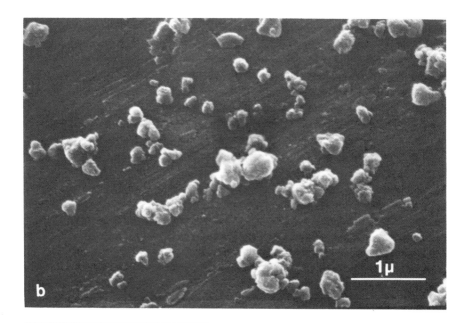

Figure 3.11. Scanning electron micrograph of (a) α-SiC, (b) β-SiC.

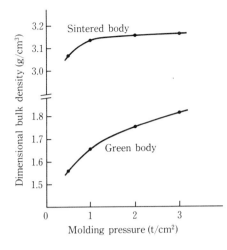

Figure 3.12. Changes in density of the green body and sintered body (pressureless sintering) according to differences in molding pressure for α-SiC fine powders.

be stated that a powder with a high pressed density is preferable for sintering, because it is easy to achieve a dense green body with it. The density of the sintered body, however, is affected by the surface activity of the particles and by several factors that control the mechanism for diffusion of the atoms during sintering. Therefore, it is necessary to evaluate many aspects of the characteristics of the powder.

Figure 3.12 presents an example of green densities before sintering and pressureless sintered densities, varying the forming pressure, for a fine powder of α-SiC.

d. SINTERING CHARACTERISTICS. The graph in Figure 3.13 shows the results of investigating the changes in sintering characteristics for three different materials, varying the sintering temperature. There are several arguments about the merits and demerits of the sintering characteristics of the α and β forms of SiC. The results, however, also differ according to the sintering conditions. Moreover, many items such as purity, particle shape, surface activity, thermal stability, and the fine structure of the sintered body remain to be investigated as factors which bear on differences in the properties of both forms. Thus, simple comparisons between the two are impossible.

e. USES AS A FUNCTIONAL FINE POWDER. Fine silicon carbide powders have long been produced as polishing powders, with those meeting JIS (Japanese Industrial Standards) Nos. 240 to 3000 and those corresponding to No. 4000 or even finer standards used for polishing and finishing quartz crystal oscillators and other high-precision applications.

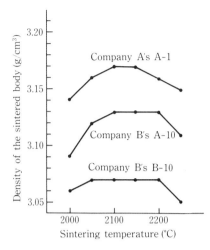

Figure 3.13. Sinterability of various types of SiC fine powders (pressureless sintering).

Applications other than sintering are also being developed, utilizing characteristics of SiC fine powders. They are being used for abrasion-resistant composite plating, fillers for powder metallurgy, heat-resistant paints, and abrasion-resistant, heat-resistant, heat-conducting fillers for resins.

2. Silicon Nitride (Si_3N_4)

Silicon nitride is another synthetic mineral which, like silicon carbide, has great promise as an engineering ceramic. Next to silicon carbide, it is the material with the strongest covalent bonding. Unlike SiC, however, it has until now been produced industrially only in small amounts as a compounding material in special high-grade refractories by nitriding metallic silicon near 1400°C. The resulting grade of silicon nitride, however, has a coarse grain size and many impurities, rendering it unsatisfactory as a material for sintering. Thus, using higher grade materials and a more advanced production technology, fine powders with outstanding characteristics are being developed (Mori, Komeya, Tsuge, and Inoue, 1982, p. 834). A variety of different approaches to its production are being pursued.

a. CRYSTALS. Silicon nitride has two crystal forms, α and β, both of which belong to the hexagonal phase. The α form is the low-temperature and the β form the high-temperature form. There has been much discussion of the crystal structure of Si_3N_4, which has not yet been fully resolved, but it is thought that the α form is apt to contain oxygen as a solid solution within the crystal. Powders principally of the α form with a slight amount of the β form as well as those of the amorphous phase are commercially available. All transform into β-Si_3N_4 after sintering.

3. Synthetic Raw Materials for Ceramics   57

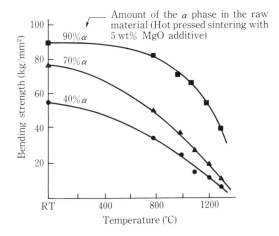

Figure 3.14.   Effect of the alpha ratio in the powdered material on strength. From Nishida et al (1973).

The α form is said to be preferable as an initial material for sintered bodies (Nishida *et al.*, 1973, p. 34; Lange, 1973, p. 518). Figure 3.14 provides information on this point. The advantage of using the α form may lie in the activated state accompanying the transformation from the α to the β form in sintering, or in the long columnar crystal structure of the sintered body when the α form is used. Some objections to those opinions remain. At present, however, the proportion of the α-form crystal in a powder, called the α-ratio, is a significant factor in evaluating the characteristics of these powders. A higher α ratio is preferred.

b. PRODUCTION METHODS.   The following methods are used in powder production.

1. Nitriding metallic silicon: The most common method, it is used by several powder manufacturers. As the reaction formula

$$3Si + 2N_2 \rightarrow Si_3N_4$$

indicates, a fine silicon powder is placed in an N_2 or NH_3 atmosphere under controlled conditions of raising the temperature from near 1200°C. After the nitriding is completed at 1400° to 1500°C, grinding and, if necessary, refining processing is performed. This reaction is fiercely exothermic, which makes it difficult to retain the high α ratio regarded as desirable in materials for sintering. The critical point is how to restrain the temperature increase and advance the nitriding reaction. In this method, the purity and particle size of the silicon are major factors controlling the purity of the fine powders of Si_3N_4. The impurities iron, aluminum, and calcium remaining in the silicon nitride powder are said to have an adverse effect on the characteristics of the sintered body. The aluminum in the silicon, the starting material, is a negative factor, while the iron acts as a positive factor, in terms of the effect on the α ratio of the powder.

2. Silica reduction: As the reaction formula

$$3SiO_2 + 6C + 2N_2 \rightarrow Si_3N_4 + 6CO$$

indicates, high-purity silica fine powder is simultaneously reduced by carbon and nitrided by heating in an N_2 atmosphere. After the remaining carbon is burned, fine Si_3N_4 powder with a high degree of purity and a high α ratio is obtained. Since this reaction is endothermic, control of the process is relatively easy; the reaction itself, however, is extremely complex. It has the advantage of making it possible to use a material with a high degree of purity and a relatively low price.

3. Nitriding silicon haloid: For the gas-phase synthesis

$$3SiCl_4 + 4NH_3 \rightarrow Si_3N_4 + 12HCl$$
$$3SiCl_4 + 16NH_3 \rightarrow Si_3N_4 + 12NH_4Cl$$

The reaction formula for imide pyrolysis (liquid phase, etc.) is

$$SiCl_4 + 6NH_3 \rightarrow Si(NH)_2 + 4NH_4Cl$$
$$3Si(NH)_2 \rightarrow Si_3N_4 + 2NH_3$$

The gas-phase synthesis principally causes $SiCl_4$ or other similar gases to react with NH_3 at 1000 to 1200°C, yielding an amorphous Si_3N_4, from which the α-form crystal is obtained after heat treatment. Removal of the remaining chlorine is necessary in this case (Maruyama, 1982, p. 40).

The imide pyrolysis method has the reaction occur at normal or even lower temperatures in an organic solvent, yielding silicon diimide, which in turn undergoes pyrolysis in an N_2 or NH_3 atmosphere at 1200° to 1350°C, promoting the removal of the ammonia and crystallization (Maruyama, 1982, p. 40; Kasai, Tsukuma, and Tsukidate, 1979, p. 97).

These methods produce an extremely fine powder with a higher degree of purity than do those discussed earlier, but there are many peripheral problems with the technology, such as means of preventing oxidation during handling of these powders and of forming good green compacts; these aspects must be improved in the future. The cost of the initial materials is high, but, as chemical processes, they also have room for cost reduction through mass production. They will conceivably be considered among the methods for producing ceramic materials in the future. Currently, several firms in Japan and abroad are experimenting with pilot production using these methods.

c. CHARACTERISTICS OF SILICON NITRIDE POWDERS. Table 3.10 summarizes measurements of the characteristics of commercially available silicon nitride powders.

1. Chemical composition: Iron, aluminum, and calcium are the most important impurities within the powder, for they become problems with respect to reduction of strength in high-temperature regions and resistance to oxidation. All of these powders contain 1 to 3% oxygen, partly, it is thought, dissolved in the α crystals and partly existing as an oxidized layer (an SiO_2 layer) on the surface of the particles. Its effect upon the sintered body is a delicate point, but a greater oxygen content than necessary is undesirable, for it causes a decrease in high-temperature strength.

2. Particle size and particle shape: Thus far the discussion over submicrometer-sized particles which occurred for SiC has not arisen for Si_3N_4. Submicrometer powders have become readily available recently; they are produced by the silicon nitriding method. They will probably be considered more in the near future, however.

Particle shape varies, corresponding to the variety of production methods. It is not certain which shape is most desirable. It has been predicted (Haggerty and Cannon, 1980, 78-073 and 81-003; Haggerty, 1981, 82-002; Cannon et al., 1982, p. 324) that the extremely fine particles with uniform particle size and spherical shape found in some of the gas-phase-method powers are ideal, but it is difficult to evaluate the effect of particle shape. Figure 3.15 shows particle shapes in a scanning electron micrograph of powder obtained by silicon nitriding.

3. Boron Nitride (BN)

Although it cannot rival SiC and Si_3N_4 as a high-strength, abrasion-resistant mechanical material, hexagonal boron nitride (hBN) is a material displaying unique properties. Boron nitride can also be synthesized in a high-pressure phase (cubic phase), which is used as superabrasive grains, as is diamond. That phase, however, cannot be discussed here.

a. PRODUCTION. Anhydrous boric acid or borate (borax, for instance) is heated to 800° to 1000°C with ammonia, urea, or other nitrogen compounds, making the BN bond. At this juncture, there are usually several means of controlling the reaction, such as using calcium phosphate as a filler. Then the product is refined and, after further heat treatment, a

Table 3.10. Examples of Characteristics of Commercially Available Si_3N_4 Fine Powders According to Production Method

	Nitriding metallic silicon						Silica reduction	Imide pyrolysis		Gas phase
	Company A product	Company A product	Company B product	Company B product	Company C product	Company C product	Company D product[a]	Company E product[a]	Company F product[a]	Company F product[a]
Characteristic	SN-9S	SN-7S	N-2F	N-3	H 1	LC-10		TS-7	SN-502	SN-402
Nitriding ratio (%)	99	100	100	99	100					
α ratio (%)	95	86	86	90	97	95	93	90	(95) 40[b]	100[b]
Chemical composition (%)										
Free Si	0.20	0.29	0.46			0.01				
N	36.8	36.0	38.1	38.0	37.8	36.9	37.8			
O	1.73	1.72	1.45	0.57	1.29	1.54	1.8	Cl<0.1	Cl<0.04	Cl<1
C	0.39	0.18	0.27	0.20	0.38	0.15	0.9	1	1.4	2.9
Fe	0.17	0.31	0.61	0.69	0.02	0.03	0.007	<0.1	0.05	0.5
Al	0.21	0.37	0.11	0.19	0.06	0.03	0.002	<0.005	<0.002	<0.002
Ca	0.26	0.19	0.12	0.12	0.02	0.02	0.007	<0.001	<0.001	<0.001
Particle-size distribution (cumulative wt %)	5μ | 4 2 | 20 1 | 36 0.5 | 63	4 22 39 66	6 22 40 68	44μ | 25 10 | 63 2 | 78 1 | 85	5μ | 7 2 | 22 1 | 37 0.5 | 61	0 17 36 67	10μ | 2 5 | 32 2 | 98	5μ | 0 2 | 7 1 | 15		
Specific surface area (m²/g)	6.2	5.5	5.8	—	9.8	11.6	—	12	4	11

[a] According to the value given in the catalog.
[b] Amorphous.

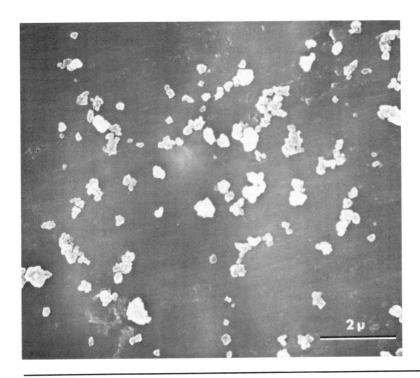

Figure 3.15. Scanning electron micrograph of Si_3N_4 fine powder (metallic silicon nitriding method).

thin-section BN powder with crystal growth comparable to graphitization is realized (Murasato, 1972, p. 243; Matsuo, 1976, p. 1066). Figure 3.16 is a transmission electron micrograph of this powder. Since the sinterability of BN is extremely poor, sintered articles are normally obtained by hot-pressing, with the addition of borates, silica, or other binders.

b. STRUCTURE AND PROPERTIES (MURASATO, 1972, P. 243: MATSUO, 1976, P. 1066). Boron nitride is similar to graphite in structure. (See Figure 3.17). Its qualities are also similar at many points. It has a quite high thermal conductivity (at high temperatures, in particular, it surpasses BeO). It also has a low thermal expansion coefficient; therefore, it shows excellent thermal shock resistance. It is stable up to 3000°C in an inert atmosphere. It can readily be processed mechanically, on a lathe for instance, and, as can be understood from its structure, it has lubricating properties.

Boron nitride also has properties not found in graphite that have led to its being valued as a unique material. In contrast to the electrical conductivity of graphite, BN has electrical insulating properties up to high

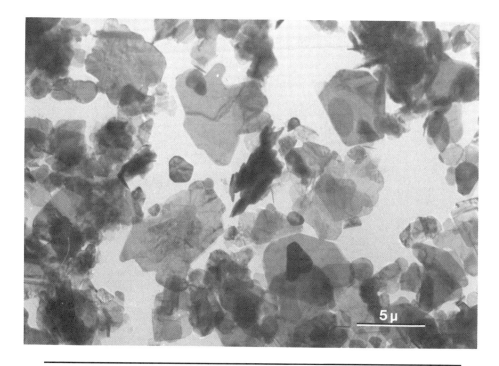

Figure 3.16. Transmission electron micrograph of BN powder.

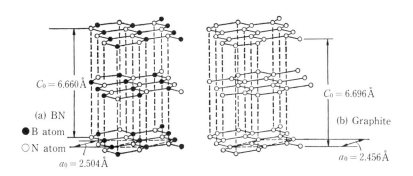

Figure 3.17. Crystal structure of BN and graphite.

temperatures. In air, graphite oxidizes at 500°C or above, but BN can be used until 900°C. Futhermore, whereas graphite reacts with metals to produce carbides, BN has outstanding resistance to corrosion, not even being wetted by most fused metals. Another distinctive feature is its white coloration.

c. USES. As a powder, an oil or water suspension type and an aerogel type of BN are available for use as lubricants and as mold-release agents. Boron nitride is also used as an additive and a filler in refractories, resins, and powder metallurgy.

As a sintered body, BN is employed industrially for parts and jigs for use with molten metals or glass. It is also used as a material for high-temperature furnaces, high-frequency electric insulating materials, and neutron absorption materials, among others.

REFERENCES

Böcker, W.; Landerfermann, H.; and Hausner, H. (1981). The influence of powder characteristics on the sintering of α-SiC. *Powder Metallurgy International*, 13, 37–39.

Cannon, W. R., et al. (1982). Sinterable ceramic powders from laser-driven reaction. I: Process description and modeling. *Journal of the American Ceramic Society*, 65, 324–330.

Enomoto, Ryo. (1982). Shōketsu genryō toshite no tankakeiso bifunmatsu [Silicon carbide power as a sintering material]. *Ceramics Japan*, 17, 828–833.

Fukuura, Isamu and Asano, Yukiyasu. (1983). Arumina seramikkusu chū no fujunbutsu no hinshitsu e no eikyō [Effects on quality of impurities in alumina ceramics]. In *Seramikku dēta bukku '83*. [Ceramic data book '83]. Tokyo: Kōgyō Seihin Gijutsu Kyōkai, pp. 116–122.

Haggerty, J. S. (1981). Sinterable powders from laser driven reactions (final report). *MIT-EL*, 82–002.

Haggerty, J. S. and Cannon, W. R. (1980). Sinterable powders from laser driven reactions (third annual report). *MIT-EL*, 81–003.

Hamano, Kenya, (1983). In *Arita Chiku Nyū Seramikku Kyōgikai Kōen yokōshū* [Proceedings of the Arita region new ceramics association]. Arita, Japan, pp. 11–16.

Hase, Teizō and Suzuki, Hiroshige. (1978). Kābon burakku no keika ni yotte chōsei sareta tankakeiso bifunmatsu no seishitsu [Properties of submicron β-SiC prepared by means of siliconization of carbon black]. *Yōgyō Kyōkaishi* [Journal of the Ceramic Society of Japan], 86, 541–546.

Hase, Teizō and Suzuki, Hiroshige. (1978). Kābon burakku no keika ni yotte chōsei sareta tankakeiso bifunmatsu no Jōatsu Shōketsusei [Sinterability of submicron β-SiC prepared by means of siliconization of carbon black]. *Yōgyo Kyōkaishi* [Journal of the Ceramic Society of Japan], 86, 606–611.

Hino, Mitsuo; Iwama, Tatsurō; Takaya, Kiyoshi; and Ichikawa, Kagetaka. (1978). Tankakeiso no tokusei ni tsuite [Characteristics of silicon carbide]. *Taikabutsu* [Refractories], 30, 115–121.

Kasai, Kiyoshi; Tsukuma, Kōji; and Tsukidate, Takaaki. (1979). Si$_3$N$_4$ no gōsei (dai 1 pō): Si(NH)$_2$ no netsubunkai kikō [The synthesis of Si$_3$N$_4$ (part 1): the mechanism of thermal decomposition of Si(NH)$_2$]. *Tōyō sōda kenkyū hōkoku* [Scientific report of the Tōyō Soda Manufacturing Co., Ltd.], 23, 97–104.

Katō, Akio; Okabe, Yasuzō; and Hōjō, Jun'ichi. (1980). Kisō hannō hō ni yoru tankakeiso chōbifuntai no gōsei e no purazuma jetto no riyō [Application of plasma jet for preparation of ultrafine silicon carbide powders by vapor phase reaction)]. *Funtai oyobi funmatsu yakin* [Journal of the Japan Society of Powder and Powder Metallurgy], 27, 32–34.

Kawamura, Takahiro and Iwama, Tatsuro. (1977/1978). Tankakeiso [Silicon carbide]. In *Seramikku dēta bukku* [Ceramic data book]. Tokyo: Kōgyō Seihin Gijutsu Kyōkai, pp. 33–39.

Kingery, W. D., Bowen, H. K.; and Uhlmann, D. R. (eds.) (1976). *Introduction to Ceramics*. New York: John Wiley & Sons.

Lange, F. F. (1973). Relation between strength, fracture energy, and microstructure of hot-pressed Si$_3$N$_4$. *Journal of the American Ceramic Society*, 56, 518–522.

Maruyama, Masaki. (1982). Kaihatsu rasshu no Si$_3$N$_4$ funmatsu gōseihō—genryō gijutsu wa shūkakuki no zenya. [The rush to develop Si$_3$N$_4$ powder synthesis—raw material techniques on the eve of harvest]. *Nikkei Mechanical*, March 29, 40–47.

Masaki, Takaki, Kobayashi, Keisuke; et al. (1981). ZrO$_2$-Y$_2$O$_3$ kei no hotto puresu shōketsutai no kikaiteki seishitsu ni tsuitte [Concerning the mechanical properties of hot-pressed ZrO$_2$-Y$_2$O$_3$]. In *Yōgyō Kyōkai nenkai kōen yokōshū* [Annual meeting of the Ceramic Society of Japan]. Tokyo: Yōgyō Kyōkai, p. 2.

Matsuo, Tadashi. (1976). Chikkahōso to junkatsusei [Boron nitride and its lubricating properties]. *Kagaku kōgyō* [Chemical Industry (Japan)], 27, 1066–1073.

May, T. C., et al. (1978). A new physical mechanism for soft errors in dynamic memories. *Ann. Proc. Reliab. Phys. [Symp.]*, 16, 33–40.

Murasato, Shigetaka. (1972). Chikkahōso (BN) no kōgyōteki riyō [The industrial applications of boron nitride (BN)]. *Ceramics Japan*, 7, 243–248.

Mori, Masaki, Komeya, Katsutoshi, Tsuge, Akihiko; and Inoue, Hiroshi. (1982). Chikka keiso funmatsu [Silicon nitride powder], *Ceramics Japan*, 17, 834–840.

Nishida, Katsutoshi, et al. (1973). Chikka keiso kaatsu shōketsutai no kyodo [Strength of hot-pressed silicon nitride]. In *Shōwa 48 nen funtai funmatsu yakin kyōkai shunki taikai kōen yokōshū* [Proceedings of the Spring 1983, meeting of the Japan Society of Powder and Powder Metallurgy]. Kyoto: Funtai Funmatsu Yakin Kyokai, pp. 34–36.

Okabe, Yasuzō, Hōjō, Jun'ichi, and Katō, Akio. (1980). SiH$_4$-CH$_4$-H$_2$ kei kisō hannō ni yoru tankakeiso funtai no gōsei [Synthesis of silicon carbide powders by a SiH$_4$-CH$_4$-H$_2$ vapor phase reaction]. *Nippon kagaku kaishi*, 188–193.

Scott, H. G. (1975). Phase relations in the zirconia-yttria system. *Journal of Materials Science*, 10, 1527–1535.

Takamiya, Yōichi. (1982). Magunesia [Synthetic magnesia]. *Ceramics Japan*, 17, 823–828.

Yōgyō kyōkai henshūiinkai kōza shōiinkai [Subcommittee on Lecture of Editorial Committee of The Ceramics Society of Japan] (ed.). (1979). *Seramikkusu no kikaiteki seishitsu* [Mechanical properties of ceramics]. Tokyo: Yōgyō Kyōkai.

4

Production Processes for Ceramics

Isamu Fukuura
NGK Spark Plug Co., Ltd.
Mizuho-ku, Nagoya 467, Japan

Tadayoshi Hirao
NGK Spark Plug Co., Ltd.,
Mizuho-ku, Nagoya 467, Japan

I. INTRODUCTION

Ceramic products are made of many different materials. (For a discussion of the materials used, see chapter 2, this volume.) These products have, moreover, a remarkable diversity of applications and forms. Not surprisingly, then, the production processes used to create them are also many and varied. Nonetheless, the basic process can be summarized in four steps: preparing the body, molding, sintering, and finishing.

As a simple list, these steps appear uncomplicated, yet what is actually entailed by each is extremely complex. For instance, consider preparing the body. The ultrafine powdered materials, on the order of micrometer diameters, used to form the body make quality control a ticklish problem. The molding stage also presents special problems: To mold ceramic bodies which are completely without elasticity requires the use of a suitable binding agent or other type of molding aid. During sintering, the molded object commonly softens and shrinks; unless appropriate measures are taken at that point, the dimensions of the product will be nonuniform and its shape deformed. The finishing stage, after sintering, also presents unusual difficulties.

As a result of these difficulties in production, the ceramic products seen up to now have largely been objects of comparatively simple form, most often used in applications where they functioned independently, rather than in combination with parts made of other materials. In recent years, however, the outstanding qualities of ceramics as a material—qualities not found in other substances—have come to the fore. Research in applications such as precision structural parts has been lively, with fierce competition to improve existing production processes or to develop entirely new ones. In addition, Japan's Ministry of International Trade and Industry (MITI) has, since 1981, commissioned private industry to launch development projects to raise the level of ceramics production technology. This enterprise is part of its Research and Development Project of Basic Technologies for Future Industries, for which over Y10 billion (about $80,000,000) in research and development funds has been allocated to stimulating the creation of the basic technologies for the next generation of industrial growth.

The specific targets for this development effort cover all four production steps: making the pulverized ceramic material, molding, sintering, and finishing. The goal in developing new technologies for handling the pulverized material includes achieving a high degree of purity and ultrafine particle size through improved methods of synthesis, pulverizing, classification and surface treatment. In molding technology, the aim is the capability to mold complex forms, with research focused on the mixing (bonding), molding, drying, and processing stages. Research on sintering techniques has aimed at attaining homogeneous, precision sintering through high pressure or continuous sintering. At the finishing step, work on processing bonding techniques focuses on accurate, swift bonding and accurate, high-strength bonding. In addition, MITI's plans include the development of assay techniques for ceramic materials and of applied technology for ceramic parts.

The achievement of these research and development goals is likely to change production processes considerably. In this chapter, however, we introduce existing production processes.

II. PREPARING THE BODY

High-performance ceramic products in general are made from a high-density sintered body with a homogeneous micro structure and few holes. To yield such high-quality products, nothing is more essential that strict selection and control of the initial materials. Ceramic bodies are composed of main constituents, crystalline structures, and auxiliary constituents that assist the sintering of the main constituents, forming grain-boundary layers. To produce a high-density sintered body, it is better to use a fine powder with a high degree of purity as the ceramic body powder. In addition, the auxiliary constituents must be evenly mixed into the main constituents. Well known in practice, this point was explained on a theoretical level by Kingery (1955) with his theory of lattice diffusion sintering and Coble (1958) with his theory of grain-boundary diffusion sintering.

A. SELECTING THE POWDERED MATERIAL

The most widespread method for producing ceramic body powder is to turn the lumps of raw materials into a fine powder by means of mechanical pulverization and classification. There are many types of pulverizers, but ball mills and jolt mills are most suitable for reaching micrometer-level grain size. A critical point concerning the pulverized material is that, while regulating homogeneous powder size is important, grain-size distribution also has a significant effect on the sintering qualities of the body. Thus, both powder size and distribution must be carefully considered. The operating conditions for the pulverizer are the first factors to be considered as affecting its performance in producing a fine, uniform powder. Given the need to consider productivity as well, however, setting optimal operating conditions is not a simple issue.

In addition, the method chosen to adjust the grain-size distribution may be to blend particles that have been classified by size. Using such mechanical means to prepare the powder, however, is excellent in terms of productivity but less desirable in terms of quality control, since it cannot completely avoid the danger of the admixture of impurities. There is also a limit to the grain size which can be achieved by mechanical means.

Further increasing the performance of ceramic products will require purer, ultrafine powdered materials. To meet those needs, materials with dramatically improved purity and grain size have been developed using a variety of synthesis methods for ceramic materials from the liquid or gas phase. These methods have been adopted by industry as well. Chapter 3 (this volume) provides a discussion of these techniques.

B. MIXING THE BODY

There are two methods, the wet and the dry, for mixing the powders, the main and auxiliary constituents, to prepare the body. In the emerging field of new ceramics, in which powders at the micrometer and submicrometer level must be handled, however, the more efficient and reliable wet method is preferred. Also, pulverizing of the original material and mixing of the body are often performed simultaneously. Water is the solvent most often used in the wet method, but for nonoxide constituents such as silicon nitride and silicon carbide, an organic solvent is used to prevent their oxidation. An additional factor is the tendency for ultrafine powders to cohere into strong aggregate particles; thus, a suitable deflocculant is often added.

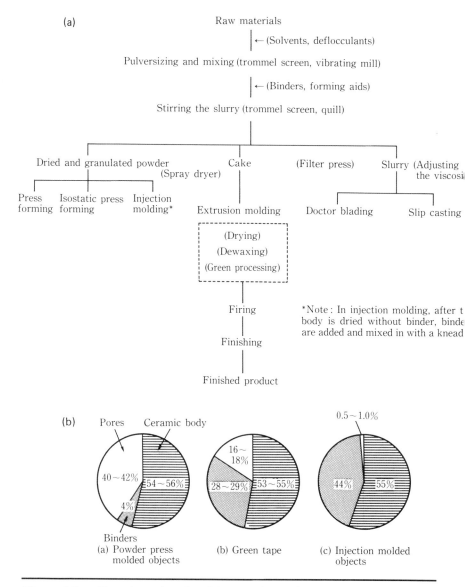

Figure 4.1. Flowchart showing (a) ceramic production processes and (b) ratio of constituents for molded objects.

A ball mill is the most commonly used mixing device. Since, however, there is the danger that particles rubbed off the ball or off the brick lining will fall into the material being mixed and contaminate it, it is preferable to use a ceramic ball and lining brick of the same material as that being handled.

The body, in slurry form after the mixing is finished, is dried, made into powder, and readied for the molding process. Or binder constituents can be added to the slurry, which, without being dried, is provided as a slurry to the molding process. Or, in slurry state, it is run through a filter press and delivered to the molding process as a conventional, moist ceramic body. A flowchart illustrating the process of preparation for these three types is given in Figure 4.1.

C. METHODS OF DRYING AND GRANULATING THE BODY

The most common means of drying the body formed by the wet method uses a spray dryer. It is not only a rational drying method but also can bring about the good body flowability desirable in powder molding. The principle is that hot air is forced into a hopper in which the slurry is being sprayed as a mist, due to centrifugal force, from disk-shaped nozzles rotating at high speed. The droplets of mist, spun by the draft of hot air, land in the bottom of the hopper as dry granules and are removed. The hot air used to dry them, now full of moisture, is disposed of from the side of the hopper bottom.

In the processing, as little solvent as possible is used in the ceramic body slurry; the selection of a deflocculant with the right viscosity for atomization is also critical. With too much solvent, granulation does not proceed successfully and many fine particles are caught up and borne away by the cyclonic wind, lowering the yields. Naturally, the heat efficiency also worsens. With the usual solvent, water, a slurry with a 25 to 45% water content is used. The granules formed by the drying and granulation are 50 to 200 μm in size, are spherical, and are high in flowability.

III. THE MOLDING PROCESS

A. MOLD PRESSING AND RUBBER PRESSING (ISOSTATIC PRESSING)

The most commonly used molding method is to pack the dry, granulated body, prepared as described above, in a mold and to pressure-form it. There are two types, mold press forming and rubber press forming (the isostatic method), in commercial use.

In mold press forming, the body is packed inside the cavity of the mold. It is then compressed by top and bottom punches. This method is the same as the molding method used in powder metallurgy. Unlike metallic powders, however, ceramic body powders are poor in hardening qualities, even under pressure. Thus, a small amount of binder is required. If the binder is added to the ceramic body slurry before it is granulated, the binder will remain as a film on the surface of the granules after the spray drying. This binder film, furthermore, preserves the strength of the granules and of the molded body after molding. If the granules are hard, they are difficult to crush, which implies excellent molding qualities. The disadvantage, however, is that, during the molding, the granules do not break down fully; the result may be gaps within the molded substance so that densification will not advance fully. If soft granules are used, however, there are fewer problems with densification, but the molding properties may be harmed by the body's adhering to the surface of the mold punch. Thus, the binder film has a considerable influence on the molding properties of the ceramic body and on its properties after sintering. Techniques for binder film molding are a critical aspect of the manufacturer's know-how.

The equipment for press forming is simple and the molding costs are low. The technique has its advantages, then, but the application of pressure in only one direction causes great friction between the powder and the mold walls, and transmission of the pressure within the powder may suffer. Those factors can, depending on the shape of the object to be molded, lead to substantial unevenness in the transmission of pressure, which can cause distortions in shape and cracks when the object is fired. Therefore, the mold press forming method is used only for molding comparatively simple shapes in which these problems are unlikely to arise. In addition, since the ceramic powders used have a high degree of hardness, it is necessary to consider the dimensional precision of the molded object and the life span of the mold, due to friction against it, in choosing press forming.

In contrast, in rubber press forming, the powdered body is packed into a rubber container, sealed, and then press-formed through the rubber membrane from all directions under normal hydrostatic pressure. Therefore, the unevenness in density and the directional quality of the pressed powder which can occur with mold press forming are unlikely to arise.

In the wet method of rubber press forming, the rubber mold, charged with ceramic powder, is hermetically sealed. Then it is immersed in a suitable oil and placed under pressure. The wet method is ideal as a method of press forming, but is weaker in terms of practical applicability.

In the dry method, the rubber mold is fixed in a high-pressure-producing device. The powder is packed in through the opening in the mold, the mold is hermetically sealed, and pressure is applied all around it, except at the opening. Although this method cannot provide perfectly even compression, it is highly suitable for use in mass production.

In a rubber press forming, also, the molded object becomes a pressed powder compact in contact with the rubber membrane, making precision

4. Production Processes for Ceramics 71

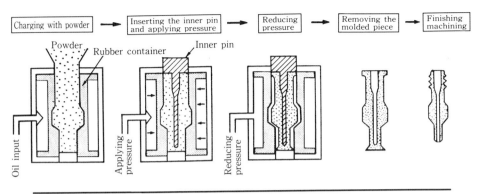

Figure 4.2. Dry process for isostatic pressing.

molding difficult to achieve. Normally, the details are finished afterward by emery polishing. In addition, in this molding method, the selection of the quality and thickness of the rubber used considerably affect the practicality of the method and the life of the mold. In terms of maximizing the life span of the rubber mold, the thicker the rubber membrane, the better. But with a thick membrane, in the pressure release process after forming, the rubber membrane has a greater righting moment, that can lead to destruction of the pressed-powder compact. In addition, it becomes more difficult to apply uniform pressure as the rubber membrane becomes thicker. An example of the use of the dry method of rubber press molding in mass production of ceramics is its use in manufacturing porcelain components for automobile spark plugs (Figure 4.2).

B. INJECTION MOLDING

In injection molding, after heat-mixing thermoplastic resins, wax, or plasticizers into the ceramic powdered body, the mixture is deaerated and formed into pellets. Next the pellets are supplied to the molding device, melted within the heating tube, then forced at high speed into a mold. The forming process and equipment used are almost identical with those used in injection molding of plastics. For ceramics, however, the formed object must be fired after the binder is extracted, a complication which has made the injection process impractical for ceramics. The range of its applications has not gone beyond thread guides for the spinning industry and some products related to electronic components.

In recent years, however, with growing interest in the new ceramics as high-strength, heat-resistant materials, serious effort has been made to develop new applications such as the ceramic gas turbine or the ceramic engine, which require many parts in complicated shapes impossible to

C. THE DOCTOR BLADE METHOD

With the type of device sketched in Figure 4.3, it is possible to coat a carrier tape or film with a slurry to a certain thickness, then strip off the carrier tape after drying and hardening, thereby producing the sheet-formed ceramic body known as green tape. By selection of the appropriate binder and plasticizer for mixing into the slurry, it is possible to give the green tape characteristics that make it easy to process by punching, lamination, and so on. In addition, a pastelike ink made of metallic powder (tungsten or molybdenum), resins, and a solvent can be printed on the green tape. If the green tape is then fired in a wet hydrogen atmosphere, the metal pattern printed on the surface of the sintered ceramic will be

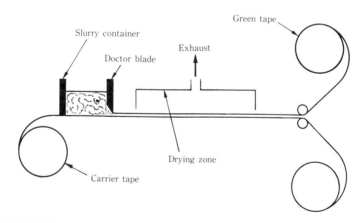

Ceramic material	Organic binder	Plasticizer	Dispersion medium
	Polyvinyl butyl resin, etc.	Phthalic ester, etc.	Ketones, alcohols, hydrocarbons, etc.
1000 parts	6-10 parts	4-8 parts	50 parts

Figure 4.3. Doctor blading.

strongly fired onto it. This phenomenon is called metallizing. The development of green tape, which is easy to mass produce, has gone hand in hand with its applications processing technology. It has been the occasion for the use of many ceramic products by the electronics industry, in electronic circuit substrates, integrated circuit packages, and multilayer substrates.

Figure 4.4 indicates the production process for one application of green tape, an integrated circuit package. A rather different use of green tape is

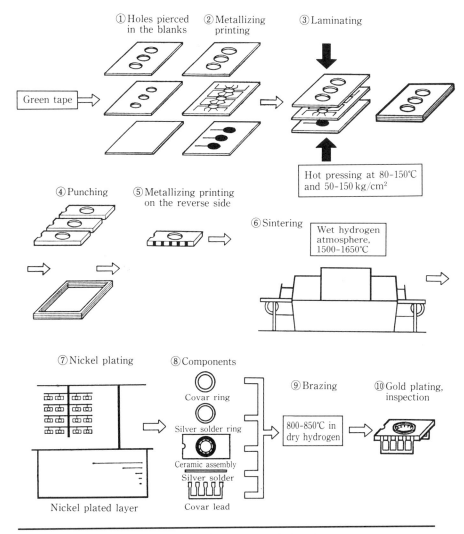

Figure 4.4. *Production process for integrated circuit packages.*

in the formation of honeycomb-structured objects. Green tape is corrugated, in a process similar to that used to make corrugated paper, and rolled up. When extrusion, a forming technique discussed below, is used, the equipment used places limits on the cell number density and size attainable. With green tape, however, it is possible to make the walls thinner. Thus, large-scale honeycomb structures can be produced with a high cell number density as well as a large aperture ratio. Uses in such areas as heat-exchange devices are being considered.

D. SLIP CASTING

Pour a slurry into a plaster of paris mold, leave it for a time, and the water content in the slurry will be absorbed by the mold. Then the excess slurry is poured off, the molded object is dried in its mold, and then, when it is dry and strong, it is removed from the mold. Control of the casting thickness is determined by the amount of time allowed to pass from the pouring of the slurry into the mold until the pouring off of the excess slurry.

Slip casting is a molding method unique to ceramics and with a long history of use. It is used for forming a wide range of objects, from analytical crucibles and boats to sanitary wares. Very thick-walled castings cannot be obtained, but quite complicated forms can be cast. Consideration is being given to using this method in the ceramic gas turbine; the concept has excited considerable interest recently, but will require development of techniques for creating the plaster of paris molds with greater precision. Thought must also be given to the life span of the plaster of paris mold.

E. EXTRUSION

A slurry has its water content squeezed out in a filter press to produce a cake, which is then inserted into a vacuum extrusion machine. The ceramic object is formed by extrusion by attaching one of a variety of nozzles to the vacuum extrusion machine, as shown in Figure 4.5. Extrusion is used to form long objects with simple cross-sectional forms, such as cylindrical or square rods. Some products are also made by machining the extruded rod form after it has dried.

As this sketch of the principal ceramic forming methods indicates, each method is suited for some applications and unsuited for others. Thus, in production planning, it is necessary to choose the appropriate forming method after careful consideration of, above all, the shape of the object to be produced, but also the delivery time, the quantity needed, the cost, and other factors. The amount and type of binder used are also largely decided by the forming method. Thus, the ultimate decision about the molding

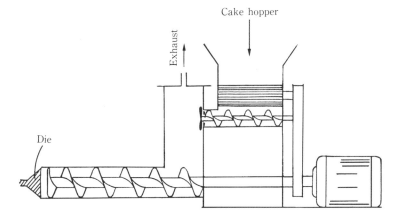

Figure 4.5. *Device for extrusion molding.*

technique used requires consideration of the characteristics of the ceramic body. Table 4.1 summarizes the distinctive features of the major methods of forming ceramics.

IV. THE FIRING PROCESS

After the binder has been removed from the molded or cast object (still an aggregate of ceramic powder), it is fired and hardened into a new state, becoming a unified, dense object. This process is called firing, and the phenomenon brought about by firing is called sintering. Firing requires heat and a device to bring that heat to bear on the molded object effectively—that is, a kiln.

There are many types of kilns, but in terms of function they may be divided into batch kilns and continuous kilns. In batch kilns, the objects to be fired are loaded in the kiln, fired, cooled, and removed; this set of operations is carried out on a single-batch basis. Thus, batch kilns are not well suited for large-scale mass production, but they are the appropriate choice for processing especially large or long objects. The batch kiln also has the advantages of being flexible in regard to firing conditions and of being relatively inexpensive to build.

The continuous kiln is used for firing high-volume products. Figure 4.6 illustrates the tunnel-shaped continuous kiln. It has three zones, for preheating, firing, and cooling. The products are loaded on carts which enter the kiln from one end, one by one, at a given interval of time. The cart advances on the rails a certain distance, then, according to the temperature

Table 4.1. Types of Molding Methods and Their Characteristics

Molding method	Molding additives (amount per 100 parts of ceramic powder)	Shape of product

Press forming	Binder:
	  Water-soluble resin ⎫
	  Emulsion ⎬ (3.0–5.0)
	  Gum arabic, etc. ⎭
	Water (0.3–1.0)
		Small, flat objects
		Various containers for electronic components
		Liners for mills

Rubber press molding (isostatic pressing)	Binder
	  Water-soluble resin ⎫
	  Emulsion ⎬ (2.0–5.0)
	  Gum arabic, etc. ⎭
	Water (0.3–1.0)
		Tubular, columnar, or spherical Forms
		Balls for grinding
		Insulators for spark plugs
		Outer tube for vacuum tubes

Injection molding	Binder
	  Thermoplastic resin ⎫
	  wax, etc. ⎬ (10.0–25.0)
	Plasticizer
	  Phthalic ester ⎫
	  Various oils, etc. ⎬ (0.5–5.0)
		Mass production of products in complex shapes
		Thread guides for spinning
		Valves
		Turbine vanes

Doctor tape method	Binder
	  Polyvinyl butyl, ⎫
	  acrylic ester, etc. ⎬ (8.0–15.0)
	Plasticizer
	  Phthalic ester, etc. (3.0–8.0)
	Solvent[a]
	  Alcohols, ketones, ⎫
	  hydrocarbons, etc. ⎬ (50)
		Thin sheets (1.5 mm or less thick and secondary processed products of such sheets
		Laminated packages
		Multilayer substrates
		Thin- and thick-film substrates

Slip casting	Binder
	  Sodium alginate ⎫
	  Methyl cellulose, etc. ⎬ (0.5–3.0)
	Water (15.0–30.0)
		Thin-walled products in irregular shapes
		Crucibles
		Thread guides
		Blades

Extrusion	Binder
	  Wax emulsion ⎫
	  Water-soluble resin ⎬ (8.0–15.0)
	Water (15.0–30.0)
		Long objects in columnar, tubular, etc, forms
		Insulator tubes
		Protective tubes
		Ceramic honeycomb
		Pipes

Hot-press molding	Normally no additives are used.	Especially objects which require high density and high strength Ceramic tools Piezoelectric substances

[a] There are also water-soluble types.

Table 4.1. (*Continued*)

Technical point	Advantages	Disadvantages

Adjustment and control of granules
Pressing conditions
  Pressure
  Pressing speed
  Duration of pressing, etc.
	Least expensive method of forming
	Readily automated
	Product can be fired directly after molding
		Density tends to be uneven in forming, limiting the sizes and shapes which can be molded

Adjustment and control of granules Pressing conditions Selecting rubber quality and thickness Receptivity to cutting and finish processing Reuse of shavings	Density of the formed object is uniform with few distortions of shape The dry method permits mass production	Many cases require external shaping and cutting after forming The life of the rubber mold is short The wet method is not commercially practicable
Kneading the body and binders Dewaxing Injection conditions Designing the mold	Can mass-produce products in complex, three-dimensional shapes Outstanding dimensional accuracy Surface finish is good	Mold costs are high, making the method unsuited to small-scale production The dewaxing process is difficult
Slurry control Preventing stretching or shrinking of the sheets Selection of the forming aid to give the material sheet characteristics such as the ability to be laminated	Outstanding productivity and good accuracy in the thickness produced A range of secondary processing of the sheets is possible No dimensionality in the shrinkage in firing	The equipment is expensive The equipment takes up a great deal of space Care must be given to explosion prevention measures and to health and safety
Slurry control Design of the plaster of paris mold The drying method	Can be produced with simple equipment Can create relatively complex shapes	Distortions frequently occur in the casting The casting process is time-consuming Many plaster of paris molds are needed, taking up space
Design of the dies Drying method	No special limitation on the length of the object to be extruded Continuous production is possible	A drying step is needed after the forming When objects with large cross sections are to be formed, the equipment is large
The equipment as a whole Choosing the appropriate pressure	The density after sintering is increased, and physical properties are improved It is possible to reduce the sintering temperature	There are major limitations on forms Apart from a few special products, lacks suitability for mass production

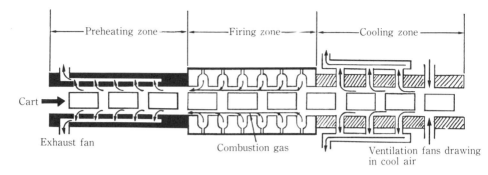

Figure 4.6. Structure of a tunnel kiln.

distribution created within the kiln, goes through the preheating, firing, and cooling processes, and is removed at the opposite end. The cooling zone is designed to cool the fired ceramics gradually; ventilation fans placed near the exit pull in air from outside to cool the products. With a tunnel kiln, great care must be taken in loading the pieces to be fired on the cart. Improper loading can produce a high-temperature gas drift phenomenon, leading to uneven firing. The kiln atmosphere is usually kept neutral, but oxidation or reduction firing can also be supported as needed. In addition, in firing such ceramics as integrated circuit packages, a tunnel kiln with flame curtain seal structures at the entrance and exit is used. These structures preserve the wet hydrogen atmosphere needed to carry out the metallizing process. Because of the high cost of the hydrogen gas used, in many cases a mixture of hydrogen with inert nitrogen gas is used.

A. REACTION SINTERING

The method known as reaction sintering is used with the nonoxides such as SiC and Si_3N_4. In the reaction sintering of silicon carbide, a mixed powder of silicon carbide powder and graphite powder is used as the body. A binder is added and, after molding, the binder is extracted. Then the molded object is heat-treated in a vacuum furnace to melt the silicon, whereupon the graphite reacts with the silicon within the object, resulting in a sintered object of silicon carbide.

Reaction sintering of silicon nitride uses metallic silicon powder as the initial material. As in SiC molding, a binder is added and then removed after forming. Then the molded object is heat-treated in an N_2 atmosphere, during which the N_2 gas, which has penetrated to every part of the metallic silicon molded object, reacts with the metallic silicon, producing Si_3N_4.

The distinctive feature of reaction-sintered ceramics is that the reactant penetrates to all parts of the porous fabric of the already formed piece and, as the product of the reaction is formed, the molded object turns into a sintered object with almost no change in dimensions. Thus, reaction sintering can produce ceramics with complicated shapes and high dimensional precision. The method does have its disadvantages, however. Since the Si vapor and the N_2 gas participating in the reaction must penetrate the formed body, it is difficult to produce a fully dense product. Moreover, in thick-walled pieces, the reaction may not proceed all the way to the interior; thus, the method is suitable only for making relatively thin-walled ceramics.

B. PRESSURE SINTERING

In the ceramics sintering reaction, raising the temperature promotes diffusion. The interior pores are pushed to the outside, and sintering accelerates. High temperatures, however, bring risk: Abnormal grain growth is more likely to occur, homogeneity of texture may be impaired, and the properties of the product may suffer.

To promote diffusion without raising the temperature too high, a method involving the application of external pressure during sintering has been developed. It is now an accepted part of the production process. In pressure sintering, the sintering is completed at relatively low temperatures without affecting the homogeneity of the texture. In addition, less sintering aid is required, and the density and quality of the resulting products are higher than those made with ordinary sintering methods.

Pressure-sintering methods include hot-press sintering and isostatic hot-press sintering. The hot-press sintering method uses a device like that illustrated in Figure 4.7. In a heated state, pressure transmission is carried out from the upper punch; the pressure produced is received by the carbon case. The restriction which applies with this equipment is that hot pressing is limited to the sintering of objects with simple shapes. If the object is too thick in the hot-pressing dimension, pressure transmission will be poor, and the product may not have the hoped-for properties. Moreover, since a carbon case is used, pressure can be applied only up to about 500 kg/cm^2. Ceramic cutting tools are an example of a product for which the hot-press sintering method is being successfully applied.

In hot isostatic pressing, the piece to be fired is first coated with a special glass; then pressure for sintering is applied to it from all directions, isostatically, by means of a high-pressure gas. Unlike hot pressing, there are no limitations on direction, and high pressure can be applied, to about 3,000 kg/cm^2. The properties of the sintered pieces thus produced are outstanding, but the method entails high operating and equipment costs. Moreover, hot isostatic pressing is still in the test stage for many applications.

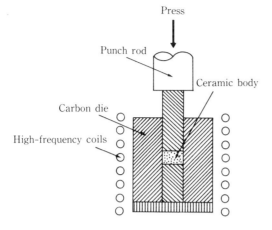

Figure 4.7. *Hot-press molding device.*

C. THE FINISHING PROCESS FOR CERAMICS

Progress in forming and sintering techniques has made special finishing unnecessary in many product areas. A rapidly growing number of fields, however, demand product precision which cannot be met by the pace of development in forming and sintering techniques. Thus, the importance of finishing, too, is growing.

The purposes of finishing can be divided into two classes: to increase the dimensional precision of the sintered product and to increase its surface smoothness. To achieve more precise dimensions, grinding with a diamond grinding wheel is used. To smooth the surface, polishing with polishing powders is used.

The features of diamond wheels depend on the type of binder in which the diamond particles are embedded. To give a general classification, there are electrodeposition grinding wheels, metal bond grinding wheels, and resinoid wheels. With electrodeposition wheels, the diamond particles are fixed through the medium of a nickel-plated layer on the base metal. Since part of the grinding particles are exposed at the surface, this type of wheel has good cuttting powers. Although it is suited to cutting materials which tend to cause loading, the finished surface is not good. Metal bond wheels bind the diamond particles to the base metal with a low-melting-point alloy in the copper–tin group. In this type of wheel the bonding strength is high, the finished surface is somewhat inferior, yet the wheel is relatively simple to use and has a long working life. It is suited for dry grinding, such as profile milling. In resinoid wheels the diamond particles are bonded by a phenolic plastic. Bond strength is weak and heat resistance is also inferior, but this type of grinding wheel is suitable when a high level of finished surface is required.

When the grinding process is actually considered, however, in addition to the selection of the type of grinding wheel, the speed of the wheel, the feed ratio, the feed speed, and a variety of other grinding conditions must be specified to attain a good result. There is no space here to touch on each condition, but a point common to all cutting of ceramics is that, in comparison with the cutting conditions for other materials, setting the wheel speed higher will produce better results, including better cutting and longer wheel life. Finally, it is becoming clear that the cutting processing method used has a considerable influence on the strength of the ceramic object. From now on it will be an important point for consideration in the processing of ceramic structural parts that require great strength.

V. ISSUES FOR THE NEAR FUTURE IN CERAMICS PRODUCTION TECHNIQUES

Progress in production techniques has two aspects. One is related to quality: to the creation of new techniques to bring about properties, forms, or densities hitherto unattainable. The other aspect concerns progress in improving productivity: speeding up and automating the production process and reducing costs. A the present time, both public research institutions and industry are energetically involved in improving quality, and considerable progress has been achieved. The question of productivity, however, has not received the attention it deserves. In particular, consideration of productivity issues for the nonoxide ceramics attracting so much attention now, including SiC and Si_3N_4, has lagged behind.

An appropriate benchmark for considering productivity issues is the speed at which metals are formed and processed. To realize the high expectations held for ceramics in many quarters, to bring about the coming of the second Stone Age, we must actively seek cooperation from experts in a wide range of fields and press on to improve both quality and productivity.

REFERENCES

Coble, R. L. (1958). Initial sintering of alumina and hematite. *Journal of the American Ceramics Society,* 41, 55–62.

Kingery, W. D. and Berg, Morris. (1955). Study of the initial states of sintering solids by viscous flow, evaporation-condensation, and self-diffusion. *Journal of Applied Physics,* 26, 1205–1212.

5

Evaluating Ceramics

Hiroshi Okuda
Japan Fine Ceramics Center
Atsuta-ku, Nagoya 456, Japan

I. INTRODUCTION

Ceramics are used in electronic, chemical, and mechanical engineering as well as in a wide range of other engineering fields, with each field demanding particular sets of properties in the ceramic objects it utilizes. Naturally, then, it is necessary to learn what the electrical, mechanical, and chemical properties of ceramics are. Methods of evaluating ceramics, in particular their mechanical and thermal properties, are the subject of this chapter.

II. MECHANICAL PROPERTIES

There are two types of ratings of the mechanical properties of ceramics: those concerned with static load, such as flexural strength, tensile strength, and compressive strength, and those concerned with dynamic load, including impact strength, friction, and abrasion. It is also possible to subdivide each type of strength into immediate breaking strength, in which the load is increased at a set load velocity or strain velocity to break the object, and delayed breaking strength, in which a certain load is applied for a long period of time, as in creep rupture strength and fatigue strength. Other measures of physical properties are also extremely important in ceramics, including hardness, the elastic constant, and fracture toughness. The most important of these factors are discussed below.

A. TESTING FLEXURAL STRENGTH

Bending tests are the most widely used of all methods for testing the strength of ceramics, since the test specimen is easy to prepare and the test itself is relatively simple. In mechanical design, however, it is tensile stress, not bending stress, that is the factor generally used; thus, a device is needed to convert bending stress in some way to tensil stress.

The bending test, in which tensile stress is induced in a plane of the test specimen is derived from the theory of the simple beam. For brittle materials such as ceramics, the strength of the material does not, however, depend only on the greatest stress it can withstand; the distribution of stresses within the test specimen has a great effect as well. Therefore, in testing flexural strength, the shape and size of the test specimens, the test equipment, and the testing method must be selected to ensure that there is as little deviation as possible from the stress distribution given in the basic theory.

Figure 5.1 illustrates the advanced ceramics flexural strength test method specified in JIS (Japanese Industrial Standard) R 1601. Flexural strength is calculated according to Eqs. (5.1) (three-point bending) and (5.2) (four-point bending), using the variables defined in Figure 5.1.

$$\sigma_{b3} = \frac{3PL}{2wt^2} \tag{5.1}$$

$$\sigma_{b4} = \frac{3PL(L - l)}{2wt^2} \tag{5.2}$$

The major issues that must be noted concerning this bending test are discussed next.

1. Influence of the Surface Condition of the Test Specimen

Usually the test specimen is ground with a diamond grinding wheel, but this surface treatment adds the danger of causing defects in the test specimen. The relationship between the grain size of the grinding wheel used and the flexural strength is given in Figure 5.2. Moreover, cracks caused by the processing may not only affect the surface but may also extend to the interior, even building up a work-altered layer in the specimen.

Figure 5.3 gives a comparison of the Weibull plots of the failure probabilities measurements for hot-pressed silicon nitride (HP Si_3N_4) in which

5. Evaluating Ceramics   85

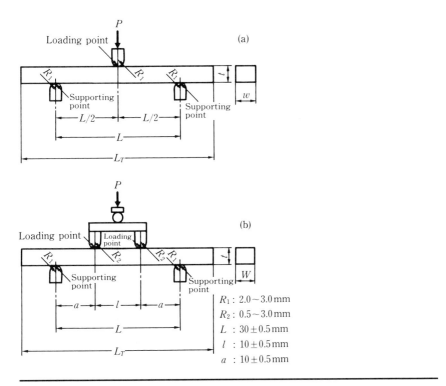

Figure 5.1. Test method for flexural strength (JIS R 1601): (a), three-point bending; (b), four-point bending.

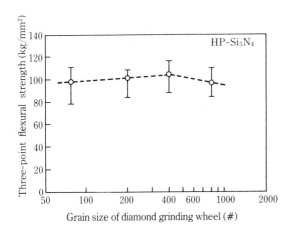

Figure 5.2. Effect of grain size of grinding wheel on the flexural strength test.

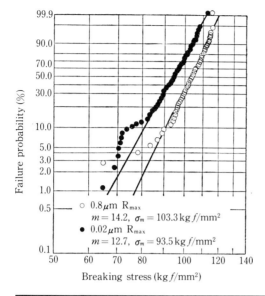

Figure 5.3. Effect of surface finishing process on flexural strength of HP Si_3N_4.

the surface 0.8 μm R_{max} has been finished using a diamond grinding wheel and then in which that surface 0.02 μm R_{max} has been further finished with a WA 60K vitrified grinding wheel. Despite a lessened roughness of the surface, an approximately 10% reduction in strength was found after the second surface finishing. This result is believed to be caused by the formation of a work-altered layer at the surface of the specimen, due to the great thermal and mechanical load placed on it by the use of WA grinding wheel, which has bad grinding qualities with respect to the extremely hard HP Si_3N_4.

The direction of the surface finishing on the test specimen is another issue. In general, when the grinding is carried out at right angles to the direction of tensile stress placed on the specimen, the specimen's strength declines.

As these results indicate, the strength of ceramic materials is considerably influenced not only by their surface roughness but also the direction in which surface grinding was applied, the type of grinding particle used, and the grinding conditions.

2. Effects of the Shape and Dimensions of the Test Specimen

Whether the specimen has a round or square cross section has an effect on the measured values for its strength. In general, test specimens which are thick near their neutral axis have greater flexural strength than do thinner ones. That is, a round rod shows greater strength than a square one, which in turn shows greater strength than an I-shaped rod (Newnham, 1975, p. 281).

For brittle materials such as ceramics, the scale effect has a considerable effect upon the strength of test specimens. The influence of size upon breakage is a question of the probability of the presence of defects; thus test specimens with larger dimensions display lower strength than do smaller pieces. Equations (5.3) and (5.4) express the relationship between scale and strength using effective volume and effective surface area.

$$\sigma_{f1} = \left(\frac{V_{E2}}{V_{E1}}\right) \frac{1}{m_{\sigma_{f2}}} \tag{5.3}$$

$$\sigma_{f1} = \left(\frac{S_{E2}}{S_{E1}}\right) \frac{1}{m_{\sigma_{f2}}} \tag{5.4}$$

where V_{E2}/V_E, is the effective volume ratio, S_{E2}/S_E, is the effective surface area ratio, and m the Weibull coefficient.

Similarly, if σ_{m3}, σ_{m4}, and σ_{mt} are set equal to the average strength under three-point and four-point bending for the test specimen of the JIS-defined shape given in Figure 5.1 and the average tensile strength for a specimen of the same dimensions, their relationships can be grasped in terms of effective volume in the following equations:

$$\frac{\sigma_{m4}}{\sigma_{m3}} = \left(\frac{V_{E3}}{V_{E4}}\right)^{1/m} = \left(\frac{3}{m+3}\right)^{1/m} \tag{5.5}$$

$$\frac{\sigma_{mt}}{\sigma_{m3}} = \left(\frac{V_{E3}}{V_{Et}}\right)^{1/m} = \left(\frac{1}{2(m+1)^2}\right)^{1/m} \tag{5.6}$$

Equations (5.5) and (5.6) apply when the breakage is caused by a defect in the interior of the test specimen. If the cause of the breakage is at the surface of the specimen, it is necessary to write the equations in terms of effective surface.

3. Flexural Strength at High Temperatures

Measured values of the strength of ceramic materials at high temperatures are essential data in the design of structural materials for high-temperature applications. Measurement in the high-temperature range of 1000°C or more, however, presents many difficulties. In particular, it is necessary to give full consideration to the effect of the hot atmosphere on the test specimen and the load jig. For atmospheric measurements, load jigs of ceramic materials—alumina, silicon nitride, or silicon carbide—are used.

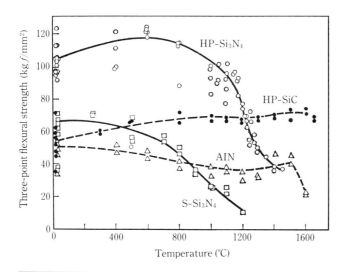

Figure 5.4. Flexural strength at high temperatures for several types of ceramics. From Okuda et al. (1983, p. 142).

For measurements in a vacuum or in an inert gas atmosphere, jigs of graphite, tungsten, or molybdenum are used. In all cases, it is necessary to bear in mind the reduction in measurement accuracy due to the form of the jig and its deformation.

Figure 5.4 presents measurements of flexural strength at high temperatures in a vacuum for four ceramic materials, HP Si_3N_4, HP SiC, AlN, and pressureless-sintered Si_3N_4. The ways in which their strength varies with temperature are distinctively different for each type of material.

B. TESTING TENSILE STRENGTH

Tensile strength is pertinent in the design of mechanical parts. Many points concerning the testing of the tensile strength of ceramic objects remain problematic, however, and the establishment of an accurate means of measuring tensile strength would be desirable. Both the conventional testing methods and those now being considered use a round rod for the test specimen. With this design, however, the concentration of stress at the point of contact with the clamps and the axle alignment have been sources of concern.

The tensile strength testing method sketched in Figure 5.5 is used for ceramics (Itō, Itō, and Sakai, 1981). The lath-shaped test specimen is held between rubber pads in a flat chuck to prevent breakage of the portion in the chuck. Figure 5.6 graphs the strength data for several types of ceramics, as measured using that method, in a Weibull distribution table.

5. Evaluating Ceramics   89

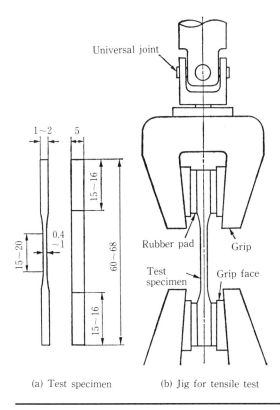

(a) Test specimen   (b) Jig for tensile test

Figure 5.5. Simple tensile test.

C. ELASTIC CONSTANTS

To discuss the mechanical properties of ceramics it is necessary to learn its elastic constants, including Young's modulus, the rigidity, and the Poisson ratio. Recently, the setting of a JIS for the test method for the elastic modulus has been under examination. The JIS proposal is that tests for both the static and dynamic elastic modulus be employed; the test method for the static elastic modulus would be the standard, and the dynamic test method would be based on it.

The test for the static elastic modulus uses bending tests for the simplicity of measurement they afford; a test specimen at least 1-mm thick, 4-mm wide, and 30- to 80-mm long is used; three-point or four-point bending load is added, the distortion measured, and the elastic modulus calculated.

For measurement of the dynamic elastic modulus, both the bending resonance method and the ultrasonic pulse method are used. In the bending resonance method, a rectangular test specimen at least 1-mm thick, 5-mm wide, and 40-mm long is placed on the support and primary resonance vibrations in the 200 to 10,000 Hz frequency range are measured. The

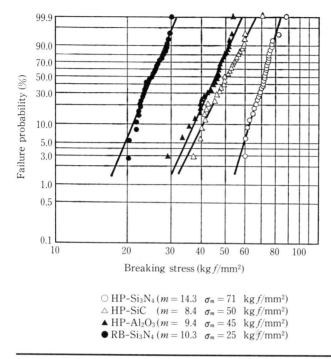

○ HP-Si$_3$N$_4$ ($m = 14.3$ $\sigma_m = 71$ kg$f$/mm^2)
△ HP-SiC ($m = 8.4$ $\sigma_m = 50$ kg$f$/mm^2)
▲ HP-Al$_2$O$_3$ ($m = 9.4$ $\sigma_m = 45$ kg$f$/mm^2)
● RB-Si$_3$N$_4$ ($m = 10.3$ $\sigma_m = 25$ kg$f$/mm^2)

Figure 5.6. *Tensile strength of ceramics.*

ultrasonic pulse method is a means of finding the sonic velocity within a ceramic object and calculating the elastic modulus. There are two versions, the reflection method and the transmission method. The test specimen is basically a square rod at least 10 mm on a side or a round column at least 10 mm in diameter. If measurements are taken by the transmission method, the test specimen is at least 40-mm long. To make the measurements, an ultrasonic pulse of basic frequency 0.2 to 30 MHz is excited, and the time needed for the ultrasonic pulse to be transmitted through the test specimen or be reflected from it is measured using measuring equipment capable of accuracy within 0.05 sec, and the elastic modulus is computed.

Table 5.1 presents the Young's modulus, the rigidity, and the Poisson ratio for several types of ceramics. The values were measured using the ultrasonic pulse method at 10 MHz.

D. HARDNESS

The hardness of ceramics is expressed as Rockwell hardness or as microhardness. In determining Rockwell hardness, a diamond indenter with a radius of curvature of 0.2 mm and a circular conic angle of 120° is used.

Table 5.1. The Elastic Constants of Ceramics

Constant	Alumina (Al_2O_3)	Reaction-sintered Si_3N_4	Hot-pressed Si_3N_4	Hot-pressed SiC
Young's modulus (10^3 kg/mm^2)	37.5	21.0	31.5	43.3
Rigidity (10^3 kg/mm^2)	15.2	8.6	12.4	18.9
Poisson's ratio	0.238	0.216	0.275	0.147

Normally, hardness is expressed by A scale Rockwell hardness (HRA) with a standard load of 10 kg and the test load 60 kg and by Rockwell superficial hardness (HR 45 N) with a standard load of 3 kg and the test load of 45 kg, as specified in JIS Z 2245 and JIS B 7726.

Microhardness tests use a conic diamond indenter. A test load of 50 to 1,000 g is pressed onto the test surface and the material's hardness is found from the size of the resulting dent; there are two such methods, Vickers hardness (Hv) and Knoop hardness (Hk). In measuring Vickers hardness, a regular pyramidal indenter with facing angles of 136° is employed. The hardness is calculated from the length of the diagonal line between the test load and the dent. For Knoop hardness, a four-sided pyramidal indenter of diamond-shaped horizontal section with vertically opposite angles of 172° 30' and 130° is used. A dent is made on the test surface, and the hardness is calculated from the relationship between the load and the surface of projection (JIS Z 2251). Measurements of the hardness of several types of ceramics are given in Table 5.2.

E. FRACTURE TOUGHNESS

Ceramics fracture when a crack occurs, due to the concentration of stresses at a minute flaw within the material or on its surface. The crack grows and

Table 5.2. The Hardness of Ceramics[a]

Hardness	Alumina (Al_2O_3)	Reaction-sintered Si_3N_4	Pressureless sintered Si_3N_4	Hot-pressed Si_3N_4	Reaction-sintered SiC	Hot-pressed SiC
HRA	92	86	91	92	92	96
HR 45 N	87	74	85			92
Hv (0.5)		1040	1460	1690	2300	2960
Hv (1)		930	1390	1650	1980	2610
Hk (0.5)		970	1360	1610	1930	2020
Hk (1)		890	1210	1460	1630	1880

[a] Okuda et al. (1983, p. 148).

produces a fracture. To make this characteristic clear, in the concepts of fracture mechanics, which begin with linear fracture mechanics carrying out two-dimensional analysis of stress, the stress intensity factor is found from analyzing the stress conditions at the tip of the crack. This analysis requires finding the fracture toughness as a distinctive value for each type of material.

The stress intensity factor K_{IC} at the critical point when the crack starts its radical growth is called the fracture toughness. If the breaking stress is σ_f and the critical crack length is a_0, the fracture toughness can be found by Eq. (5.7).

$$K_{IC} = \sigma_f a c^{1/2} Y \tag{5.7}$$

There are two approaches to measuring fracture toughness, the large-scale and the micro. In the large-scale crack method, a notch or groove is made by an accurate method in the test specimen, to create a large crack artificially. Then the fracture toughness is found by carrying out bending tests, tensile tests, and so on. In the micro method, as an extension of the test evaluation method for hardness, an indenter is pressed with a heavy load on the test surface, causing a minute crack, and the stress intensity factor is found.

In the large-scale method, a notched beam (SENB), a bending test specimen with a notch cut into it, is used. Commonly, three-point or four-point bending tests are performed on it. The notch width in this case has a substantial effect that can give rise to error; the wider the notch, the lower the stress at its tip and the higher the material's fracture toughness. Additional methods include the double cantilever (DCB) method [see Figure 5.7(a)] and the double torsion (DT) method [Figure 5.7(b)] in which the growth rate of the crack is measured and the fracture toughness calculated. In the DCB and DT methods, the relationship between the crack growth rate and the stress intensity factor is found. These data are used in calculating the life of ceramics.

In the surface microflaw method (SMF), one of the micro methods for measuring fracture toughness, a semicircular crack is made with a Vickers or Knoop indenter on the surface of the bending test specimen receiving the tensile stress. Then four-point flexural strength is found and the stress intensity factor is calculated. In the microindentation method (MI), a Vickers indenter is pressed onto the test substance. The load creates a dent, and four cracks are produced on diagonal lines crossing at the center of that dent, The fracture toughness is found from the relationships of the load, dent, and cracks. In the SMF method, when the indentor is applied, a residual stress is created, necessitating the removal of the surface layer to eliminate the residual stress. In the MI method, the dent made on the test

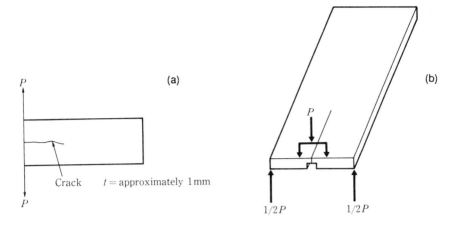

Figure 5.7. Fracture toughness test methods: (a) double cantilever test specimen; (b) double torsion method.

specimen is only on the order of 100 μm; thus this method is excellent for relative evaluation of small test specimens and for quality control. Table 5.3 presents the fracture toughness results for HP Si_3N_4 made using the SENB, DCB, and MI methods.

Table 5.3. Fracture Toughness of HP Si_3N_4

Measurement method	Fracture toughness $(MN/m^{3/2})$
SENB	7.3–10.2
DCB	6.3
MI	6.5–8.0

III. THERMAL PROPERTIES

Data on the thermal capacity, thermal conductivity, thermal expansion coefficient, and other thermal properties of ceramics are of great engineering significance. Accurate values for these properties are especially essential for ceramics used under high-temperature conditions. In addition, resistance to heat stress, that is, thermal shock resistance, is also an important factor.

A. HEAT CAPACITY

Heat capacity C_P is calculated according to Eq. (5.8) from the temperature increase ΔT when a given energy Q is applied to a test material m (mol):

$$C_P = \frac{Q}{m \, \Delta T} \tag{5.8}$$

The methods of measurement can be classified according to the heating method used: the pulse method, the adiabatic method, and the drop method. The adiabatic method can be used to measure heat capacity with high accuracy at low temperatures; the drop method is a relatively simple method for measuring at 500 K or less; to the present day, these two methods have been the mainstream of measurement for chemical thermodynamics. Recently, however, pulse methods based on laser flashes or other sources have come into use.

The variations of thermal capacity with temperature for three types of ceramics (MgO, SiC, and AIN) are given in Figure 5.8. In general, ceramics are porous, and thermal capacity declines with greater porosity.

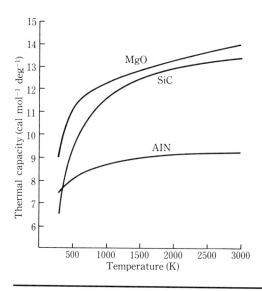

Figure 5.8. *Changes in thermal capacity of ceramic according to temperature. From Okuda et al. (1983, p. 159).*

B. THERMAL CONDUCTIVITY

The amount of thermal conduction within a solid depends on the number of carriers which carry the heat energy and the degree of resistance to their movement within the solid. The most important carriers are lattice vibrations (phonons). In ceramics, phonon transmission is dominant; in this case, the thermal conductivity depends on the temperature. The phonons are, however, dispersed by lattice defects and by the lattice vibrations themselves. In particular, the degree of dispersion increases for lattice vibrations in proportion to $1/T$. Thus, the thermal conductivity shows a maximum at a certain temperature and then declines on the higher temperature side of that point in proportion to $1/T$. The maximum value for the thermal conductivity of Al_2O_3, MgO, BeO, and other ceramics is a considerably low temperature—room temperature or below.

Measuring methods for thermal conductivity can be roughly divided into the static method, which finds the steady state thermal conductivity, and the nonstatic method, which finds the rate of thermal diffusion. Each method has its advantages and disadvantages. Thus, it is necessary to choose a measuring method according to the size of the test specimen, its properties, the degree of thermal conductivity, and the temperature range for the measurements.

The thermal conductivity of the major types of ceramics is graphed in Figure 5.9. Thermal conductivity declines as porosity increases. Also, impurities (including additives) have a significant effect upon thermal conductivity. For instance, AlN is a quite good thermal conductor but,

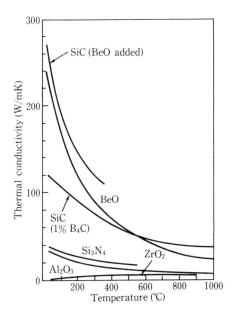

Figure 5.9. Thermal conductivity of ceramics. From Okuda et al. (1983, p. 160).

when oxides have been added to it as a sintering aid, the thermal conductivity of sintered AlN is drastically reduced. In contrast, sintered SiC to which a minute amount of BeO has been added has a thermal conductivity several times higher than normal SiC and about 1.2 times higher than SiC with metallic aluminum (Ura and Nakamura, 1982, p. 55).

C. THERMAL EXPANSION COEFFICIENT

When ceramics are used under conditions of thermal stress, their thermal expansion coefficient is one of their more important properties. The thermal expansion coefficient (α) differs according to the direction of the crystal axis, except for glass and cubic crystals. Since in general ceramics are polycrystalline substances, even if the thermal expansion of the constitutive crystal is anisotropic, the substance as a whole will exhibit an averaged isotropic expansion. In such a case, the average linear expansion coefficient for the polycrystalline substance will prove to be one-third of the coefficient of volume expansion of the constituent crystal. Therefore, when the thermal expansion coefficient for a specific axis shows a negative value, the average linear coefficient of expanison for the polycrystalline substance will be extremely small. Such ceramics include cordierite (the MAS group) and lithium aluminum silicate (the LAS group).

The thermal expansion ratio of crystals depends largely on the crystal structure and the bond strength. Therefore, a list of thermal expansion ratios would, broadly speaking, include covalent bonds, ion bonds, metallic bonds, and molecular bonds, in ascending order. Also, when the arrangement of the constituents of the crystal structure is close-packed or very nearly so (as in MgO, BeO, or Al_2O_3), the substance will have a large thermal expansion coefficient. Conversely, when the crystal structure is complex (as in zircon or mullite), the thermal expansion coefficient will be relatively small. Figure 5.10 presents the thermal expansion coefficients for major ceramic materials.

D. RESISTANCE TO THERMAL SHOCK

Resistance to thermal shock concerns the birth of cracks and the extension of already existing cracks. The former is called thermal shock fracture resistance, the latter thermal shock damage resistance.

Thermal shock fracture resistance is an issue where the occurrence of a crack itself would be a major defect, for instance, in electrical components or gas turbines. These parts suffer sudden heating and cooling so that thermal stresses are set up by nonuniform distribution of temperatures within the part. The question is whether those stresses exceed the breaking strength of the parts.

Thermal shock damage resistance is an issue for many refractory materials, such as refractory brick. These materials are used under highly

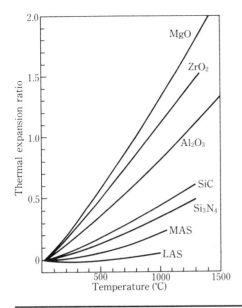

Figure 5.10. Thermal expansion ratio for several ceramics. From Okuda et al. (1983, p. 161).

demanding conditions in which the production of cracks is unavoidable. Thus, more than the occurrence of the cracks, the growth of the cracks which have already appeared, the lowering of strength, the partial exfoliation of the material, and its collapse are the problems. In such cases thermal shock damage resistance is the appropriate property used to indicate thermal shock resistance. In this case, the lower the loss of strength and the smaller the volume of exfoliation under given conditions of rapid heating and cooling, the greater the resistance to thermal shock damage. This property is unrelated to whether cracks occur.

The coefficient of thermal shock resistance, which expresses these types of resistance to thermal shock, using the thermal characteristics and mechanical properties of the material, is expressed as shown in Table 5.4 in response to a variety of thermal conditions and evaluating conditions (Hasselman, 1970, p. 1033).

The occurrence of cracks or their expansion due to thermal shock reduces the breaking strength of the material. Making a connection between the fracture behavior of ceramics and the degree of thermal shock, an attempt has been made to analyze uniformly the occurrence of fractures (the occurrence of cracks) and the scale of the damage (the expansion of cracks). The relationship between the breaking strength of the material and the degree of thermal shock (the thermal shock temperature difference, ΔT) is graphed in Figure 5.11 (Hasselman, 1969, p. 600).

Table 5.4. *Formulas for Calculating Coefficients of Thermal Shock Resistance*[a]

Coefficient of thermal shock resistance	Thermal shock condition
Thermal shock fracture resistance coefficients	
$R = \sigma(1 - \mu)/E \cdot \alpha$	Drastically rapid cooling
$R' = R \cdot k$	Gradual heating and cooling
$R'' = R'/\rho C$	Heating and cooling at a fixed speed
$R_{cy} = \sigma(1 - \mu)/\alpha \cdot \eta$	Stress relaxation due to creep
$R_{rad} = (R'/\varepsilon)^{1/4}$	Radiant heating
Thermal shock damage resistance coefficients	
$R'''' = E \cdot \gamma/\sigma^2(1 - \mu)$	Comparison of degree of expansion of cracks
$R''' = \gamma/\sigma^2(1 - \mu)$	Comparison of degree of expansion of cracks
Other coefficients	
$R_{st} = (\gamma/E \cdot \alpha^2)^{1/2}$	Crack stability
$R'_{st} = R_{st} \cdot k$	Crack stability

[a] σ = strength, μ = Poisson's ratio, E = Young's modulus, α = thermal expansion coefficient, k = thermal conductivity, ρ = density, C = specific heat, η = apparent viscosity constant, ε = average blackness of the test material surface and the radiant surface, γ = surface breaking energy.

In Range I of Figure 5.11, the strength is constant and no fractures occur. In Range II, with temperature difference ΔTc, a sudden discontinuous decline in strength occurs, with a rapid expansion of cracks. In Range III, once again strength and cracks are stable, but upon entering Range IV the strength gradually declines and cracks also expand continuously. The form of this typical temperature-difference curve for breaking strength and thermal shock varies depending upon the size of the initial

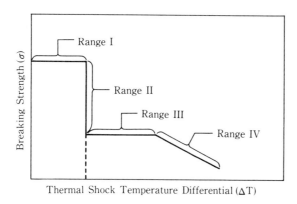

Figure 5.11. *Relationship between breaking strength (σ) and the thermal shock temperature differential (ΔT).*

cracks. The actual strength-reduction curve does not, moreover, always take this form. The phenomenon of strength reduction by thermal shock in brittle materials such as ceramics can, however, be explained for the most part with this curve.

REFERENCES

Hasselman, D. P. H. (1969). Unified theory of thermal shock fracture initiation and crack propagation in brittle ceramics. *Journal of the American Ceramic Society,* 52, 600–604.

Hasselman, D. P. H. (1970). Thermal stress resistance parameters for brittle refractory ceramics: a compendium. *American Ceramic Society Bulletin,* 49, 1033–1037.

Itō, Shōji; Itō, Masaru; and Sakai, Seisuke. (1981). Seramikkusu no hippari kyōdo [Tensile strength of ceramics]. *Dai 31 kai meikō-shi kenkyū happyōkai yokōshū,* 61–62.

Newnham, R. C. (1975). Strength tests for brittle materials. *Proceedings of the British Ceramic Society,* 25, 281–293.

Okuda, Hiroshi, et al. (1983). Enjiniaringu seramikkusu no seishitsu [Properties of engineering ceramics]. *Enjiniaringu seramikkusu [Engineering Ceramics],* 37, p. 142.

Ura, Mitsuru and Nakamura, Kōsuke. (1982). Kōnetsu dendōsei SiC kiban zairyo [SiC substrate materials with high thermal conductivity]. *Denshi zairyō,* 21, 55–59.

II
PROPERTIES AND APPLICATIONS OF CERAMICS

6

Electrical and Electronic Properties

Kikuo Wakino
Murata Manufacturing Company Limited
Nagaoka-kyo-city 617, Japan

I. INTRODUCTION

In 1887, S. F. B. Morse invented the telegraph. The growing system of telegraph lines soon created a great demand for pin-type insulators of a uniform size and shape. Then, as electrical power began to be widely used, demand for electrical porcelain for use in high-voltage, large-scale electrical power transmission also developed.

Electronic ceramics are ceramics produced for their special electrical and electronic properties. The rapid development of electronic and electrical technology has led to an ongoing series of great advances in electronic ceramics. The wide range of applications developed for these ceramics covers everything from the giant unimolded porcelain tube insulators, up to 12 long, used in ultrahigh voltage (1,000 kV) electrical power substations to thin insulators used in tunneling junction devices that have the mini-scale electrical power consumption of 10^{-6} watt.

II. INSULATING PROPERTIES

A. INSULATION MATERIALS FOR HIGH-VOLTAGE APPLICATIONS

Insulators for high-voltage power transmission lines maintain nearly absolute stability and reliability over a long period of service while withstanding harsh environmental conditions—atmosphere, sunlight, heat, wind, rain, and snow—as well as mechanical stress.

Cristobalite porcelain contains 15–40% crystalline cristobalite (SiO_2). Used in the production of large insulators in which strength is important, it is well suited for this application because of its high flexural strength (1,200–1,400 kg/cm^2), good formability, and ease of firing (Fujimura, 1978, p. 326).

Porcelain which contains a high proportion of alumina has a high firing temperature and presents difficulties in producing large insulators. However, it has a lower thermal expansion coefficient than cristobalite porcelain, and thus is better able to withstand temperature changes. It is, furthermore, characteristic of alumina porcelain that a crack is unlikely to spread. Thus, if a crack develops as a result of mechanical percussive shock, only a small piece is likely to break off.

Spark plugs used in internal combustion engines in automobiles and other vehicles not only must handle high voltages (10–25 kV) but are also exposed to extremely high temperatures and pressures, on the order of 2500°C and 50 kg/cm^2 or more, during ignition and combustion in the power stroke of the cycle. Then, in the next instant, the air-gas mixture is drawn into the cylinder in the intake stroke, resulting in sudden cooling. This cycle of operations adds up to very harsh conditions under which spark plugs must operate.

Spark plugs usually are made from a 90–96% alumina porcelain with samll amounts of CaO, MgO, SiO_2 added to lower the sintering temperature and thus facilitate the mass production process (Katō, 1978, p. 35). The higher the alumina content of alumina porcelain, however, the greater the thermal conductivity and the mechanical strength. Thus, research efforts are continuing to improve the quality of this ceramic material.

Boron nitride (BN) is light in weight and has excellent electrical insulating properties. Moreover, it resists high temperatures up to 2200°C, and its thermal shock resistance is unusually high. For this reason, it offers considerable promise as a high-temperature insulating material in such applications as furnace materials for magnets hydrodynamic electric generators (Matsui, 1980, p. 21).

B. SUBSTRATES

A glance inside a modern electronic product reveals a strikingly exact expression of the term *miniaturization*. That is, the same capabilities are being produced in a device that is smaller, lighter, thinner, easier to carry and to handle, and takes up less space than earlier versions. Overall, the miniaturized device uses up fewer resources, is more energy efficient, costs less, and is usually more reliable. Moreover, the savings in weight and space through miniaturization are a stimulus to creating devices with even higher performance.

Ultraminiaturization and integration of parts are advancing rapidly because of the demand for and benefits of small size, light weight, and high performance. With the development of parts and integrated circuits the size of grains of rice, and then large-scale integrated circuits, hybrid integrated circuits, and high-density mounting, whole systems are moving toward even higher performance in ever smaller devices.

As electronic circuits reach higher levels of integration, a need has emerged for a precision substrate with outstanding insulating and heat-conducting qualities that also permits direct mounting of integrated circuit chips and other components.

The conditions required in a substrate are

1. As high a thermal conductivity as possible
2. Superior insulation resistance, low dielectric loss, and high withstanding voltage
3. Chemical stability and low sodium ion content
4. Flat, smooth surface
5. Minimum emission of radiation, for example α rays, which can cause soft error in the large-scale integrated circuit
6. Ability to withstand heat treatment in a vacuum and other manufacturing processes
7. Superior mechanical strength without deformation and degeneration
8. Low cost

Table 6.1 presents the characteristics of a group of substrate materials. The thermal conductivity of beryllia ceramics is an order of magnitude greater than that of alumina ceramics, making it an excellent material for subtrates, for transistor packages, and for electronic circuits that give off heat. Handling the poisonous powdered BeO is dangerous, however. Therefore, fewer manufacturers will now undertake its production and processing, and alumina substrates now dominate this application field.

The thermal conductivity of single-crystal Al_2O_3 is 0.42 J/cm·s·K, so that the higher the purity of alumina ceramics, and the closer the ceramic material approximates its theoretical density, the greater the thermal conductivity that may be obtained.

Table 6.1. Comparison of Properties of Substrate Materials

Material	Thermal conductivity (J/cm·s·K)	Thermal expansion coefficient ($\times 10^{-6}$/°C)	Resistance ($\Omega \cdot$cm)	Permittivity 1 MHz
Sapphire Al_2O_3	0.42	7.3 (//)	$>10^{14}$	11.5(//)
Alumina Al_2O_3 99.5%	0.31	6.8	$>10^{14}$	9.8
Alumina Al_2O_3 96%	0.21	6.7	$>10^{14}$	9.0
Beryllia BeO 99.5%	2.40	7.7	$>10^{14}$	6.7
Zircon $ZrO_2 \cdot SiO_2$	0.05	4.3	$>10^{14}$	8.8
Mullite $3Al_2O_3 \cdot 2SiO_2$	0.04	4.0	$>10^{14}$	6.5
Steatite $MgO \cdot SiO_2$	0.025	6.9	$>10^{14}$	6.0
Forsterite $2MgO \cdot SiO_2$	0.034	10.0	$>10^{14}$	6.0
Quartz glass SiO_2	0.013	0.54	$>10^{14}$	3.6
Boron nitride BN	0.57	−0.7	$>10^{14}$	4.0
Silicon nitride Si_3N_4	0.13	2.8	$>10^{14}$	9.4
Silicon carbide SiC	2.7	3.7	—	—
HITACERAM SC-101 SiC-BeO	2.7	3.7	$>4 \times 10^{13}$	40
Silicon crystal Si	1.3	3.5–4.0	2.3×10^4	12
Copper Cu	3.9	16.8	1.7×10^{-6}	—
Aluminum Al	2.3	23.0	2.7×10^{-6}	—
Iron Fe	0.75	11.7	9.7×10^{-6}	—

6. Electrical and Electronic Properties 107

To obtain the required surface flatness and smoothness, it is usually necessary to polish and/or glaze the surface of the porcelain. It would be highly advantageous to industry, however, if substrates with a smooth, flat surface could be obtained directly out of the kiln.

Fine-grained alumina (FGA) substrates, developed by Fujitsu, are composed of 0.1%–0.4% wt% MgO, 0.001%–0.10% wt% Cr_2O_3, and at least 95.5% wt% Al_2O_3. They are made into a thin slab by the doctor blade method, then fired. Straight out of the kiln, the resulting material has a surface roughness of 0.02–0.05 μRa (Niwa and Murakawa, 1976, p. 292; Niwa, Anzai, Hashimoto, Yokoyama, 1978).

HITACERAM SC-101, developed by Hitachi, is made from α-SiC powder with an average grain size of 2 μm, with 2 wt% BeO added. This powder is then sintered by vacuum hot pressing. The substance has a thermal conductivity surpassing that of BeO, and its thermal expansion coefficient is close to that of a silicon chip, an advantage in applications as a substrate to which components are directly mounted.

The added BeO first acts as a sintering accelerator and hastens the densification of the SiC. The BeO barely dissolves into the particles of SiC. It is, instead, distributed in the grain boundaries in the matrix state, increasing the thermal conductivity at the grain boundaries, while at the same time producing superior insulation. This structure, with electrical conductivity in the interior of the SiC grains and insulating properties due to the grain boundaries, is, however, identical with the grain-boundary barrier structure described below. As a result, compared to BeO alone, its apparent electrical permittivity is high and its resistance to high voltage is low. Furthermore, Maxwell-Wagner dispersion increases the apparent dielectric loss.

These factors have meant that its uses are somewhat restricted. What has proved of great interest, however, is that, by using a combination of materials, it has proved possible to develop a ceramic product with characteristics that had not been obtainable in existing materials.

Silicon nitride blocks alkali ions effectively and is highly resistant to humidity. It is obtained by using a plasma reaction from SiH_4-NH_3 or SiH_4-N_2 gases under low pressure, and then causing deposition of the material by the activated plasma gases at 200° to 300°C. Because of its properties, it is used as a protective surface coating on integrated circuit and large-scale integrated circuit chips after they have been mounted on a substrate (Shōno, 1980).

In recent years, both the processing speeds and capacities of computers have increased radically. At the same time, computers have become much smaller. This development has made it necessary to miniaturize the active area and the circuit lengths down to the order of a micrometer. Speed will also continue to increase: access time for bipolar logic large-scale integrated circuit gates has reached 0.5 nsec or less.

Higher operating times cannot be attained if the conventional mounting method, in which parts are mounted on a board in a plane and the boards are joined with connectors at the edge of the board, is used.

Moreover, if high-density mounting is attempted with large-scale integrated circuit chips mounted on a flat substrate, overcrowding or accumulation of the conductive paths is unavoidable. To respond to this problem, mounting technology has been making rapid advances by expanding vertically, producing modules stacked in three dimensions, rather than using planar substrates alone. As a result, multilayered substrates have drawn increasing attention.

Figure 6.1 shows the structure of a multilayered ceramic substrate for the thermal conductivity module introduced in 1980 by IBM in its 3081 processor (Clark and Hill, 1980, p. 89; *Nikkei Electronics*, 1981, p. 118). The circuits are printed on the alumina green sheet with molybdenum metal paste, and then the modules are stacked and cofired into one unit. Metallization of the molybdenum creates the circuits on the alumina layers themselves. In addition, the circuits on each alumina layer are vertically connected through holes in the substrates.

In general, the many circuit lines that emerge from large-scale integrated circuit chip are taken into the surface layer of the multilayer substrate. Then a portion of them are returned to the surface. This arrangement gives flexibility since, using the terminals exposed on the surface, circuits may be changed or added to.

Multiple circuits stretched out horizontally are built into the signal layer, near the middle layer. A cooling housing is also built in and, since the chip is cooled by its water jacket, its temperature during operation can be kept below 85°C.

The greater the advances made in the degree of integration and the denser the mounting of components, the less we can neglect the problem of dealing with the heat that is generated. The goal is to develop a high-speed device that generates very little heat.

GaAs substrates, whose electron mobility within the crystal is five times greater than that of silicon, will become practical in ultrahigh-speed integrated circuits. No matter what the material used, there is a tendency to increase the number of gates per chip. That trend implies that both the planning of circuits that can take advantage of the functions of these large-scale integrated circuits and the ensuring of circuit density will become even more important.

Electronics are finding applications in a host of fields, and the miniaturization and performance enhancement of these electronic devices is the focus of intensive research. The development of multi-layer substrates for a variety of practical applications is a central part of this effort.

The molybdenum and tungsten used as conductors inside ceramic substrates have a resistance ten times higher than that of gold. Thus, there is the disagreeable possibility that they will not be able to respond to the

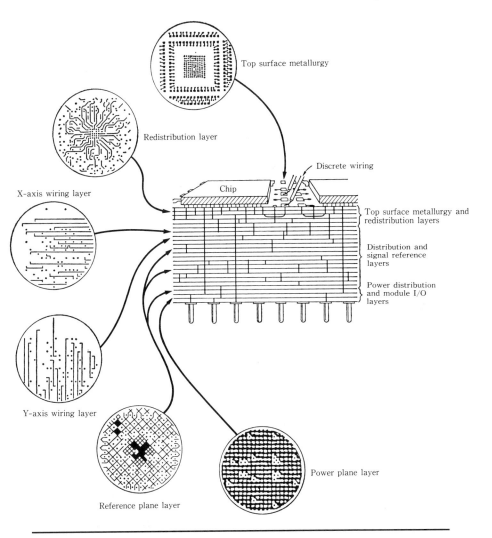

Figure 6.1. Example of multilayer ceramic circuit substrate. From Clark and Hill (1980). Copyright © 1980 IEEE.

increased speed of integrated circuits. It is likely that the development and design of shorter circuits and the effort to find a loss-free conductor will be vigorously pursued research goals.

III. SEMICONDUCTORS

When carriers, such as electrons and positive holes, are loosely bound to the crystal lattice, electrical conduction can occur. Ceramics formed from crystals with electrical conduction based on these kinds of carriers show semiconductive characteristics.

Since, however, grain boundaries exist in ceramics, they manifest different characteristics from single crystals, revealing a complex diversity of behavior that may not be seen in single crystals.

We can divide semiconductive ceramics into two large groups: (1) those that suppress the characteristics of the grain boundaries and bring out the properties of the crystal as much as possible and (2) conversely, those that exploit the peculiarities derived from the existence of the grain boundaries. Some examples of the first type include negative temperature coefficient thermistors, transparent thin films (SnO_2, In_2O_3, etc.), electrodes for electrochemicals, and resistors. The second group are called barrier-layer or boundary-layer ceramics. Some examples include positive temperature coefficient thermistors, varistors, boundary-layer capacitors, and several kinds of sensors.

A. NEGATIVE TEMPERATURE COEFFICIENT THERMISTORS

The temperature dependence of resistance of a negative temperature coefficient semiconductor crystal is expressed by the following equation:

$$R = R_0 \exp B(1/T - 1/T_0) \tag{6.1}$$

In this equation, R is the resistance at the reference temperature $T_0(K)$ and B is the thermistor constant (K). The resistance R decreases exponentially as the temperature increases.

Since negative temperature coefficient thermistor exploits the intrinsic nature of the crystals, it is necessary to choose a material for the thermistor that will not change electrical conduction properties due to adsorption of O_2 at the surface or at the grain boundaries of the ceramic. Consequently, MnO, CoO, NiO, and so on, are widely used as p-type semiconductors.

Semiconductor materials are selected on the basis of not being subject to crystal transitions or of having a high transition temperature, for stability over long periods of use.

Since materials used in temperature measuring instruments require a high degree of reliability, they are, in addition, aged for stabilization.

Silicon carbide has excellent heat resistance at high temperatures, as is obvious from its use in heating elements of electric furnaces. An SiC thin film formed by RF sputtering on the surface of, for instance, an alumina substrate can be used as negative temperature coefficient thermistor in a wide temperature range ($-100°$ to $+450°C$) (Wasa and Tohda, 1978, p. 444). These thermistors have already found application in temperature sensors in household appliances such as ovens and burners (Nagai, Yamamoto, and Kobayashi, 1980, p. 403).

B. POSITIVE TEMPERATURE COEFFICIENT THERMISTORS

Replacing the Ba^{2+} in $BaTiO_3$, which is ferroelectric, with an ion of about the same ionic radius with a higher valence, such as La^{3+}, Ce^{3+}, Nd^{3+}, or replacing the Ti^{4+} with Nb^{5+}, Ta^{5+}, or W^{6+}, produces semiconductivity. If the material is then fired in a nonreducing atmosphere or heat-treated, it will display a remarkably large positive temperature coefficient of resistance at temperatures at or above the Curie point. This kind of semiconductor is called a barium titanate semiconductor, or a positive temperature coefficient thermistor.

In ordinary ferroelectric $BaTiO_3$ ceramics, replacement by Sr, Pb, Ca, Zr, or Sn (called *shifters*) shifts the Curie point, and each additive also affects the electrical permittivity and the temperature characteristics of the material. Exactly the same phenomenon is responsible for changing the electrical resistivity and its temperature dependence. Figure 6.2 shows the typical effects on resistivities produced by each additive.

A certain amount of oxygen is necessary for the positive temperature coefficient to be apparent. Moreover, when ohmic contact is set up with a single crystal, the positive temperature coefficient cannot be observed. It is assumed from these phenomena that the positive temperature coefficient is caused by the existence of grain boundaries (including the cases in which nonohmic contact is set up on the ceramic surface) in the semiconductive ceramic.

That is, a trap level is formed by the oxygen adsorbed at the grain boundaries. This potential barrier raises the resistance of the ceramic. But,

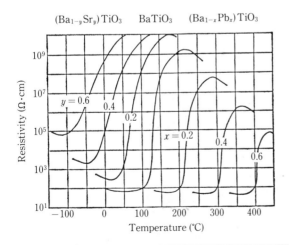

Figure 6.2. Temperature vs. resistance characteristics of barium titanate semiconductors.

below the Curie point, spontaneous polarization increases the electric field inside the ferroelectric ceramic, making the barrier low in some places and facilitating the flow of electric current even at low voltages.

Using a $BaTiO_3$ semiconductor as an example, if we apply an alternating voltage of 100 V, initially the resistance is low, so that a large current will flow. Simultaneously, however, the semiconductor generates heat. As the temperature rises steeply, the current decreases precipitately over time. Figure 6.3 shows an example of this time-current characteristic.

This feature, an alternating current decreasing within a short time, has already been put to use in the automatic degaussing circuits for color television picture tubes.

The shadow mask in a picture tube is exposed to terrestrial magnetism, and as a result becomes magnetized itself. This magnetization causes the electron beam to bend, which in turn causes color distortions. For this reason, a positive temperature coefficient thermistor is connected in serial

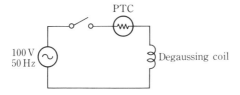

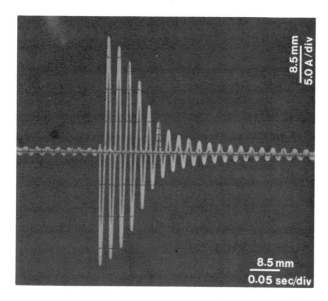

Figure 6.3. Current-time characteristics of the positive temperature coefficient thermistor.

6. Electrical and Electronic Properties   113

with the degaussing coil that is wrapped around the outside of the shadow mask. Thus, for the first one or two seconds after the power is turned on, current flows through the degaussing coil and always brings the shadow mask back to its normal state.

When current flows through a positive temperature coefficient thermistor, heat is generated and the temperature increases. However, once the Curie point is passed and the resistance suddenly increases, the current decreases and, therefore, the temperature ceases to rise. Finally, the device reaches a state of equilibrium, maintaining a set temperature. If we force the temperature down, the current will increase and the temperature will once more rise to its original level. Thus, a positive temperature coefficient thermistor, aside from acting as a heater element, can function as both as temperature controller and a temperature sensor.

A harmonica- or honeycomb-type positive temperature coefficient thermistor is often used for such things as the heating element in electric futon driers.[1] The special characteristics of this device are that, if the amount of air blown through increases, the current also increases and the temperature drops slightly and, conversely, if the air stops flowing, the electrical power is immediately reduced, eliminating the danger of overheating or of the heating element's burning out. Because of its safety and ease of use, this type of appliance has become quite popular. It is easy to imagine many similar applications.

C. VARISTORS AND BOUNDARY-LAYER CAPACITORS

In ceramics with a structure such that the conductive crystal grains are surrounded by thin insulating grain boundaries, if the potential barrier at the grain boundaries declines or suddenly yields due to the applied voltage, the ceramic is said to exhibit voltage dependence.

A resistor whose resistance changes according to the applied voltage is called a varistor.

Since the insulating layer that is distributed in matrix form in the boundary layer is extremely thin, even if the electric permittivity is small, the ceramic material exhibits a large capacitance.

When voltage is applied to the ceramic, the boundary dielectric layer is revealed to have extremely high field strength, due to the fact that the ceramic grains are semiconductors. Thus, the capacitance also shows voltage dependence. A device of this type that uses changes in capacitance according to voltage is called a varicap.

---

[1] Futon, the traditional Japanese quilt-like bedding, are normally kept folded in a closet and taken out and spread on the floor at night. They are usually hung out to air before being folded and put away, but during periods of inclement weather, people often use electric futon driers, which blow warm air through the quilts.

Of course a device for which only its large capacitance is relevant is called a capacitor, but since the device described above is based on a semiconducting ceramic, it is referred to as a semiconductor ceramic capacitor.

Sintered SiC has long been used as a varistor in relay spark suppressors, telephone automatic gain controls, and lightning arresters. The distinctive feature of this type of varistor is that the resistance changes relatively smoothly under a wide range of voltages.

A sintered body of ZnO with an admixture of Bi_2O_3 also shows a small nonlinearity of voltage-current characteristics. But if CoO, NiO, MnO, and Sb_2O_3 are added, these additives localize in the grain boundaries and form p-type boundary layers. These then enhance the nonlinearity of the voltage-current characteristics of this ZnO ceramic.

The current-voltage characteristics of a varistor is expressed by the following equation:

$$I = (V/C)^\alpha \tag{6.2}$$

where I is the current, V the voltage, and C the varistor constant; α represents the order of nonlinearity, and it can reach as high as 60 (Masuyama and Matsuoka, 1972, p. 27).

The characteristics of the varistor may be adjusted by controlling the thickness of the ceramic and the grain size. Other n-type ceramic semiconductors, such as $SrTiO_3$, $BaTiO_3$, TiO_2, SnO_2, Fe_3O_4, and so on, can also be used to make varistors, following the same principles.

Initially, $BaTiO_3$ ceramics were used for these applications, but in recent years $SrTiO_3$ ceramics have been developed, largely replacing the $BaTiO_3$ materials.

When fired in a reducing atmosphere, $SrTiO_3$ show low resistivity, about 10^{-1} $\Omega \cdot cm$, one-tenth that of $BaTiO_3$ ceramics. The dispersion frequency of $SrTiO_3$ ceramics reaches 10^9 Hz, which is superior to that of $BaTiO_3$ ceramics, at 10^8 Hz.

The problem with $SrTiO_3$ is that its relative permittivity of approximately 300 is quite small when compared with approximately 1,100 for $BaTiO_3$. However, recently developed materials have an apparent relative permittivity of 200,000, and tan δ < 1% at the working voltage of 12 V. These properties result from their extremely thin and uniform boundary dielectric layer, on the order of 0.05–0.1 μm per grain boundary.

IV. HIGH-CONDUCTIVITY CERAMICS

The vanadium oxides such as VO_2, V_2O_3, and VO undergo phase transition at the characteristic temperature T, where the electrical conductivity changes drastically and reversibly from semiconductive to metallic (Ōshima, 1973, p. 23).

For example, VO_2 has a rutile crystal structure with the phase transition at 60°C. At this temperature, the unit cell of the crystal expands about 1–2%, and the electrical conductivity increases to as much as 10^5 times higher. Above that temperature, VO_2 shows metallic conductivity.

Vanadium sesquioxide, which has a corundum structure with a phase transition at −95°C, exhibits a similar conductivity change of about 10^7 times that of its previous phase.

These materials have already been applied in a negative temperature coefficient thermistor with quickly changing characteristics.

Another interesting material is $(V_{1-x}Cr_x)_2O_3$ in which part of the vanadium ions are replaced with chromium ions. A crystal with composition $x = 0.004$ shows a jump in resistivity from 3×10^{-3} to 3×10^{-1} $\Omega \cdot cm$ at 80°C (Perkins et al., 1982, p. 225). These large positive changes in resistance are also observed in the ceramic materials.

These materials are used as sensors in applications such as overcurrent or overheating protectors.

V. SUPERCONDUCTORS

There is a class of substances called superconductors in which electrical resistance drops to zero when they are chilled to near absolute zero. Electrons in a superconductor at extremely low temperatures form pairs called Cooper pairs. They then undergo Bose-Einstein condensation to a single energy state and become able to move through the solid without being scattered.

In a superconductor, electrical loss and heat generation are zero, thus permitting a very large current to flow. Consequently, a large current in a coil can continue to flow indefinitely while generating a strong magnetic field. Since an extremely small-scale superconductor coil can show one of two states, depending on whether there is a current flowing or not, they can be used as computer memories.

Among the basic characteristics of superconducting materials, there are three critical values: T_c, the critical temperature; H_{c_2}, the upper critical magnetic field; and J_c, the critical electric current density. As a material will enter its superconducting state only below these critical values, the higher these values are, the easier it is to put the material to use, and the greater its value.

Table 6.2 shows the critical temperatures T_c for several superconducting substances. Research and discoveries in the new materials are constantly raising this value but, at present, the substance with the highest critical temperature is the compound Nb_3Ge, type A15, with a T_c of 23 K. Incidentally, the substance with the highest upper critical magnetic field, H_{c_2}, is the

Table 6.2. Superconductor Materials and Their Characteristics

Substance	Critical temperature I_c (K)	Critical magnetic field Hc_2(T, 4.2 K)	Structure
Boiling point of neon	27.10		
Nb_3Ge	23.2	—	A15
$Nb_3Al_{0.8}Ge_{0.2}$	20.7	41	A15
Nb_3Ga	20.3	34	A15
Boiling point of hydrogen	20.28		
$Nb_3Al_{0.95}Be_{0.05}$	19.6	—	A15
Nb_3Al	18.9	32	A15
Nb_3Sn	18.3	22	A15
V_3Ga	16.5	22	A15
NbN	15.7	13	B1
$PbMo_6S_8$	15.6	60	Cheverel
$LiTi_2O_4$	13.7	—	Spinel
$BaPb_{0.7}Bi_{0.3}O_3$	13	—	Perovskite
Nb-Zr	10.8	9.1	A2
Nb-Ti	9.3	12	A2
Nb	9.2	2.0	—
Pb	7.2	0.8	—
V	5.3	1.0	—
Boiling point of helium	4.25		
$SrTiO_{3-x}$	0.55	—	Perovskite

Cheverel structure of the compound $Pb_4Mo_6S_8$, with an H_{c_2} of 60 T, equivalent to 10 gauss at 4.2 K).

Until now, superconductors have found their most outstanding applications as magnets and in electric power transmission in such large-scale projects as high-energy accelerators, electric generators, energy storage devices, nuclear fusion reactors, and linear motor cars.

Superconductors have in recent years attracted attention in new applications in the field of electronics. A superconductor, whose resistance is zero, can transmit electric power at frequencies up to 10^{12} Hz without loss and can transmit a high-speed pulse to the next stage without distortion (Ōya and Onodera, 1980, p. 377). Researchers have also noticed that materials in the superconducting state have a remarkably high sensitivity to magnetic field and generate very little noise.

Thus, practical applications for superconductors include coaxial cables for communications, superconducting antennas, Josephson devices, and tunnel junction devices. Josephson devices have already found applications in detection of feeble signals (magnetic fields) and voltage standardization, but the area that has attracted the most interest is their application in Josephson tunnel junction device computers (Hasuo, 1981).

The switching time of such tunnel junction devices is on the order of 10^{-12} sec. In addition, the amount of heat generated by the device is about

10^{-6} watt. This means that they work very swiftly and consume very little electric power. Josephson tunnel junction devices have an operational speed about 50 times faster and a power consumption 1/1000 that of conventional silicon devices. These characteristics are sufficient to allow a very high degree of integration, making possible the realization of the super-small, super-powerful fifth-generation computer. Such a device would require electric power for the refrigeration necessary to maintain the temperature at that of liquid helium. But even allowing for that, the power consumption would be low, and the air conditioning now used for large computers would be unnecessary.

One element of a Josephson tunnel junction device consists of two superconducting films with an insulating film sandwiched between. In 1982, IBM developed the heavy quasi-particle injection tunnel effect device, whose structure consists of three superconducting films with insulating films between them. If a tiny current is made to flow in the middle superconducting layer, the balance in the lattice is broken and the device shows an electrical amplification effect similar to that of a transistor or triode.

In these miniature devices, the superconductors are formed in extremely thin films using physical vapor deposition and chemical vapor deposition techniques. Accordingly, appropriate materials for handling and processing by thin-film techniques are being sought. In addition, even if we design for zero thermal stress at ordinary temperatures, the stresses at the extremely low operating temperatures are likely to be much larger. Further, it may be necessary to raise the temperature to room temperature for inspection and then return it again to the extremely low operating temperatures. This temperature cycle is also likely to produce stress. One of the researcher's tasks is to minimize any deterioration due to this temperature cycle.

In 1964, the perovskite form of $SrTiO_{3-x}$, which can be obtained by strong reduction in a hydrogen atmosphere, was found to be a superconductor. This material displays superconductivity in only a narrow range of carrier densities, between 10^{19} and 10^{21}, making its handling quite delicate. The maximum critical temperature T_c is obtained when the carrier density is 10^{20}, but that temperature is only 0.55 K, making the material still less practical.

The compound $Ba(Pb_{1-x}Bi_x)O_3$ shows superconducting properties when its composition is $0.05 \leq x \leq 0.3$. When $x = 0.3$, that is, for $BaPb_{0.7}Bi_{0.3}O_3$, the maximum T_c is 13 K. In addition, this superconductor is an oxide ceramic processed in air to near 1000°C. Studies of these materials for Josephson devices have been carried out at NTT's electrical communication laboratories and other research institutions (Suzuki, Enomoto, Murakami, and Inamura, 1982, p. 1622).

In ceramics, it is easy to form an extremely thin insulating film of a homogeneous substance, which is similar to the body material, between its grain boundaries and at its surface. Being able to form these structures of homogeneous substances means that, when the device is operated at extremely low temperatures, very little thermal stress will be generated. As a result, there should be a low incidence of failure.

Nearly all of the superconducting materials hitherto discovered have required chilling by liquid helium (4.2 K). However, the utility of superconductors will increase tremendously when liquid hydrogen (20.2 K) or liquid nitrogen (77 K) can be used. Hydrogen and nitrogen are much cheaper than helium, and the refrigeration equipment needed is much simpler.

VI. IONIC CONDUCTION

A. STABILIZED ZIRCONIA

Zirconia (ZrO_2) has a very high melting point, 2,700°C, and excellent resistance to heat and corrosion (Andō and Ōishi, 1982, p. 412). But it goes through a series of phase transitions. It is monoclinic at normal temperatures, tetragonal at 1000–1150°C and above, and cubic at 2370°C and above. These transitions are accompanied by large changes in volume.

In other words, when zirconia is heated to about 1100°C, it shrinks by 7.4%; when heated to 2370°C, it shrinks by 2.4%. Conversely, when it is cooled from a high temperature, it expands by 2.4% at 2370°C and by 7.4% at about 900°C. This phenomenon means that, if zirconia were used as a refractory material, the product would be cracked by repeated heating and cooling. Thus, zirconia is not practical in such applications.

Stabilized zirconia, which was developed by dissolving a low valence oxide such as Y_2O_3, MgO, or CaO in zirconia, maintains the high-temperature phase of zirconia, that is, the cubic phase, down to room temperature. As a result, sudden structural transformations are eliminated, and stabilized zirconia is unusually stable even under cyclic temperature change conditions.

Stabilized zirconia with 15 mol% CaO added ($Ca_{0.15}Zr_{0.85}O_{1.85}$) has 0.15 molecules of oxygen vacancies. The oxygen ions from these oxygen vacancies exchange electrical charges and diffuse from the higher to the lower density side. This effect causes electrical conductivity.

Electrons, however, hardly work as carriers in these crystals. Therefore, unlike ordinary semiconductor ceramics, the electrical conductivity of stabilized zirconia is stable, withstanding heat or reduction.

Oxygen molecules do not pass through stabilized zirconia, but the oxygen ions that carry the electrical charge do. It thus acts as a permeable membrane. When a difference in the oxygen partial pressure is given, the oxygen ions with negative electrical charge move from the side where the oxygen partial pressure is high to where it is low. The high-pressure side then becomes electrically positive and the low-pressure side becomes electrically negative, thus setting up a difference in electrical potential.

The value of that difference in electrical potential, when the transference number of the electrolyte is 1, is expressed by the Nernst equation:

$$\Delta \mathscr{E} = \frac{RT}{n\mathscr{F}} \ln \frac{PO_2(\mathrm{I})}{PO_2(\mathrm{II})} \tag{6.3}$$

In Eq. (6.3), R is the gas constant, $\mathscr{F}$ is the Faraday constant, and n is the number of electrons involved in oxidation-reduction reactions.

Accordingly, the above equation may be approximated by the following:

$$\Delta \mathscr{E} = 0.0496T \log [PO_2(\mathrm{I})/PO_2(\mathrm{II})] \tag{6.4}$$

If the oxygen partial pressure on one side $PO_2(\mathrm{I})$ is known, the partial pressure on the other side $PO_2(\mathrm{II})$ can be determined by measuring the difference in electrical potential and the temperature T (K).

Because stabilized zirconia is stable under high temperatures up to about 1600°C and also is chemically stable, it is used in oxygen sensors that must be used under very severe conditions, such as in blast furnaces or in automobile exhaust equipment.

In particular, in automobiles and other motor vehicles, it is of the utmost importance to regulate the fuel–air ratio to optimize fuel combustion and reduce exhaust fume emissions. An oxygen sensor is necessary to control these processes.

Another application of stabilized zirconia is as the solid electrolyte of a fuel cell. One side is exposed to oxygen or air, and the other side is put in contact with carbon monoxide, methane, or natural gas, so-called fuel gases, which have lower oxygen partial pressure. When heated to about 700–800°C, due to the difference in concentration, the oxygen ions move to the side with the lower oxygen partial pressure, producing an electromotive force.

For example, if one side is in contact with oxygen, and the other side with hydrogen, an electromotive force of about 1.2 V will be generated. The oxygen ions that have moved react with the hydrogen and form water, and are then removed. Thus, the oxygen partial pressure on the hydrogen side is always maintained at a low value, and the reaction proceeds continuously. In other words, as long as hydrogen and oxygen are supplied, electrical power may be drawn off.

This kind of electrical generating device is called a solid-electrolyte fuel cell, or just fuel cell. Unlike conventional electrical power generation, the chemical energy of the fuel is not converted into electricity by combustion. Rather, the chemical energy is converted directly into electricity, resulting in high levels of efficiency (60–70%). Theoretically, 100% conversion efficiency is possible. Fuel cells have other advantages over conventional methods. For example, they produce no pollution, go out of order infrequently, and can be used to provide power for individual dwellings.

B. BETA-ALUMINA

Beta-alumina is a compound of a monovalent metal oxide and alumina (Wen, 1980, p. 208). The composition can be shown in the following formula: $R_2O \cdot nAl_2O_3$, where n may be between 5 and 11, and R is Na, K, Rb, Ag, and so on. Ideally, when n is 11, this compound should be called β-alumina and, when n is 5, β''-alumina. But both share a structure consisting of alumina stacked into multiple layers with a monovalent metal oxide. Figure 6.4 shows the structure of β-alumina (Yamaguchi and Suzuki, 1968, p. 93).

The alkali metal ions can move easily within the [R-O] face that the monovalent metal oxide ions occupy, resulting in ionic conduction.

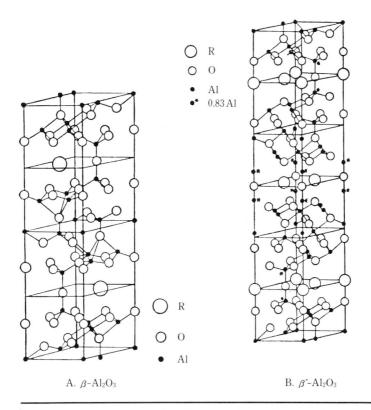

Figure 6.4. Structure of β-alumina. From Yamaguchi and Suzuki (1968).

A device which uses sodium β-alumina at an operating temperature of 300°C and above as the permeable membrane for fused sodium and sodium polysulfide (Na_2S_x) is called a sodium-sulfur cell. The metallic sodium ions move within the β-alumina and react with the Na_2S_x, discharge electricity, and become sulfides. Thus, the sodium side becomes electrically negative, and the Na_2S_x side becomes electrically positive, generating an electromotive force of about 2.08 volts.

As the reaction proceeds, the Na_2S_x gradually changes to Na_2S_3, which results in a decrease in the electromotive force. At this point it is charged with the sodium side positive and the Na_2S_x side negative, and the sodium ions move back through the permeable membrane and are turned back into metallic sodium.

This type of cell is known as a secondary cell, and its energy storage, density per unit of weight is remarkably high. The energy density of an ordinary lead-acid battery is 30–50 WH/kg. In contrast, the energy density of the Na–S cell is over 300 WH/kg. Moreover, since the action of the battery does not involve electrolysis in water, no gas is generated, and the cell may be hermetically sealed. Sealing is necessary, in fact, to prevent the electrodes from being exposed to air. Research is proceeding on the use of these cells for electrical storage and for batteries for electrical vehicles whose efficiency would rival that of gasoline engines.

C. THREE-DIMENSIONAL STRUCTURED CONDUCTORS

Because of the layered structure of β-alumina, its electrical conductivity, thermal expansion, and mechanical strength are anisotropic. Even if the β-alumina is made into a polycrystal, the anisotropic characteristics are retained in micro scale and deterioration will often occur as a result of a localized concentration of current when electricity is passed through the alumina. (Miyayama, Koumoto, and Yanagida, 1982, p. 16). Therefore, research is continuing into homogeneously dispersed three-dimensional conductors, which have pseudoisotropic electrical conduction channels, for use in large-capacity batteries.

The compound $Na_{1+x}Zr_2Si_xP_{3-x}O_{12}$, a solid solution of $NaZr_2(PO_4)_3$ and $Na_4Zr_2(SiO_4)_3$, has a monoclinic crystalline structure and shows a high level of ionic conductivity when x is between 1.8 and 2.2. When x is 2 (that is, $Na_3Zr_2Si_2PO_{12}$), the conductivity is at its maximum value of $3 \times 10^{-1} \, \Omega^{-1} \cdot cm^{-1}$ at 300°C. Named NASICON, this material is extremely stable even when exposed to molten sodium. Whereas β-alumina is sintered at 1600° to 1700°C, NASICON's sintering temperature is considerably lower, only 1200°C.

One line of research involves replacing the Zr^{4+} in NASICON with another ion. One example, $Na_3Hf_2Si_2PO_{12}$, exceeds the conductivity of NASICON at temperatures below 200°C (Cava, Vogel, and Johnson,

1982). In addition, researchers have developed $Na_5GaSi_4O_{12}$ and other compounds which show the same degree of sodium ion conductivity as NASICON.

REFERENCES

Andō, Ken and Ōishi, Yasumichi. (1982). ZrO_2 koyōtaikei no sō heikō to seibun ion no kakusan [Phase equilibrium and diffusion of constituent ions in systems of zirconia solid solution]. *Seramikkusu*, 17, 412–421.

Cava, R. J.; Vogel, E. M.; and Johnson, David W. (1982). Effect of homovalent framework cation substitution on the sodium ion conductivity in $Na_3Zr_2Si_2PO_{12}$. *Journal of the American Ceramic Society*, 65, (9), C–157.

Clark, Bernard T. and Hill, Yates M. (1980). IBM multichip multilayer ceramic modules for LSI chips: design for performance and density. *IEEE Transactions on Components, Hybrids, and Manufacturing Technology CHMT-3*, 1, 89–93.

Fujimura, Tetsuo. (1978). Gaishi yō jiki to sono tekiyō ni tsuite [Ceramics for insulators and their applications]. *Seramikku dēta bukku*, 78, 326–331.

Hasuo, Shinya. (1981). Josephson devices for computer applications [II]. *Journal of Institute of Electronics and Communication Engineers of Japan*, 64, 622–628.

Katō, Norio. (1978). Ceramic insulators for spark plugs. *Erekutoroniku seramikkusu*, 9, (Autumn), 35–39.

Masuyama, Takeshi and Matsuoka, Michio. (1972). Zinc oxide varistor. *Erekutoroniku seramikkusu*, 3, (7), 27–31.

Matsui, Fujio, (1980). Boron nitride. *Kōgyō zairyō*, 28, (5), 21–25, 52.

Miyayama, Masaru; Koumoto, Kunihiko; and Yanagida, Hiroaki. (1982). Denchi yō kotai denkaishitsu seramikkusu [Solid electrolytic ceramics for batteries]. *Erekutoroniku seramikkusu*, 13, (Winter), 16–21.

Nagai, Takeshi; Yamamoto, Kazushi; and Kobayashi, Ikuo. (1980). SiC thin-film thermistor for household products. *National Technical Report*, 26, 403–412.

Nikkei Electronics. (1981). 7–20, 118–144.

Niwa, Gen' ichi; Anzai, Yoshiharu; Hashimoto, Kaoru; and Yokoyama, Hiroshi. (1978). Tokkyo Shō 53–120 [Japan patent publication 53–120]. Japan. Nihon koku tokkyochō. *Tokkyo Kōhō*, pp. 229–235.

Niwa, Gen'ichi amd Murakawa, Kyōhei. (1976). Atarashii hybriddo IC yōarumina kiban [New alumina substrates for hybrid ICs]. *Seramikku deeta bukku*, 7, 292–296.

Ōshima, Hirotoshi. (1973). Preparations and characteristics of vanadium oxides. *Erekutoroniku seramikkusu*, Vol 4, no. 9, 23–29.

Ōya, Gin'ichirō and Onodera, Yukata. (1980). Superconducting materials. *Journal of Institute of Electronics and Communication Engineers of Japan*, 63, 377–384.

Perkins, R. S. et al. (1982). A new PTC resistor for power applications. *IEEE Transactions on Components, Hybrids, and Manufacturing Technology CHMT-5*, 2, 225–230.

Shōno, Katsufusa. (1980). ChōLSI jidai no handōtai gijutsu 100 shū [Collection of 100 semiconductor techniques for the VLSI age]. Supplement to *Electronics*,

April Oomu sha erekutoronikusu bunko, 12 [Oomu sha electronics library, vol. 12]. Tokyo: Oomu sha.

Suzuki, Minoru; Enomoto, Youichi; Murakami, Toshiaki; and Inamura, Takahiro. (1982). Preparation and properties of superconducting $BaPb_{1-x}B_xO_3$ thin films by sputtering. *Journal of Applied Physics*, 53, 1622–1630.

Ura, Mitsuru and Nakamura, Kōsuke. (1982). Kōnetsu dendōsei SiC kiban zairyō [High thermal conductive SiC substrate material]. *Denshi zairyō*, 21, (9), 55–59, 79.

Wasa, Kiyotaka and Tohda, Takao. (1978). SiC thin-firm thermistor. *National Technical Report*, 24, 444–452.

Wen, Ting-Lian. (1981). Kotai denkaishitsu toshite no β-alumina [β-alumina as a solid electrolyte]. *Fain seramikkusu*, 1, 208–215.

Yamaguchi, Goro and Suzuki, Kazutaka. (1968). On the structure of alkali polyaluminates. *Bulletin of the Chemical Society of Japan*, 41, 93–98.

7
Magnetic Properties

Teitaro Hiraga
TDK Corporation
Chuo-ku, Tokyo 103, Japan

I. INTRODUCTION

According to tradition, the strange power of lodestones to attract iron was discovered in approximately 600 B.C. This discovery was the first event in the recognition of the magnetic properties of ceramics. Lodestone, also known as magnetite, has the chemical formula $FeO \cdot Fe_2O_3$. Magnetic oxides with Fe_2O_3 as their principal constituent are called ferrite. Thus, the ancient discovery of the strange properties of lodestone had significance for the existence of iron ferrite.

Scientific research on ferrite was begun by pioneering scientists in the mid-nineteenth century. In fact, two Japanese scientists, Dr. Katō Yogorō and Dr. Takei Takeshi, took the initiative in conducting serious research orientated to industrial applications (Katō and Takei, 1932). Their series of research results on Cu ferrite and Co ferrite in the years beginning in 1932 became the nucleus and motive force which, as is well known, led to the world's first application of ferrite on a commercial basis.

Subsequently, J. L. Snoek and his colleagues at N. V. Philips Gloeilampenfabrieken published systematic fundamental research on ferrite (Snoek, 1947), and Louis Néel of France published his theory of ferrimagnetism (Néel, 1948). These publications swept ferrite into the mainstream of worldwide research on magnetism and established its position as a basic magnetic material, a position it retains to this day.

Apart from the ferrites, that is, oxide magnetic materials with Fe_2O_3 as their principal constituent, the only other oxide magnetic material is CrO_2. Therefore, in this chapter, only the ferrite group of ceramic magnetic materials is discussed.

II. FERRITE: AN OXIDE MAGNETIC MATERIAL

A. THE ORIGIN OF MAGNETISM

Atoms are characterized by the electron orbital motion around the nucleus. As the theory of electromagnetism shows, a magnetic field is included externally by the current loops. The electron itself, however, is also spinning as it travels its orbit, and that induces magnetism due to an equivalent current loop. Therefore, conceptually an atom acts upon the outside as an apparent combination of two types of small magnets.

For actual materials, however, the situation is not so simple. Ceramics, for instance, have a complex crystalline form, so that, depending on the kinds of atoms and ions forming it and the crystalline structure, a more complex magnetism may occur externally.

Ferrite is an ionic crystal formed from oxygen ions (anions) and metallic elements, principally iron ions (cations). It exists in a number of typical crystalline forms. For instance, the most common form of ferrite is a spinel structure. It is formed of 32 oxygen ions and 24 metal ions (mainly iron), interacting to generate magnetism externally. In this case, the oxygen ions, which act as the crystal skeleton, perform an extremely important function in aligning the electron spins of the metal ions. For a constituent element with a large spin magnetic moment, the iron group transition elements such as Cr, Mn, Fe, Co, and Ni, which have $3d$ orbitals, are preferable. In addition, the crystals as a whole are controlled by ferrite's characteristically strong magnetism, that is, by ferrimagnetism. That has contributed to the development of ferrite as an industrial material and to its systematic study by the scholarly world. This achievement was reflected in the 1970 Nobel prize for physics.

B. TYPES OF FERRITE

First, it is necessary to explain what ferrite means. *Ferrite* is a name given to magnetic oxides which have ferric oxide (Fe_2O_3) as their principal component. A number of such ferric oxide compounds are known to exist. They are found in the spinel structure mentioned above and in others, including hexagonal, garnet, and orthoferrite structures.

Table 7.1 presents the most important types of simple ferrite (Gorter, 1955) and their basic characteristics. In particular, $ZnFe_2O_4$ has the useful

Table 7.1. Types of Ferrites

Composition	Saturation magnetization (Gauss)	Curie point (°C)	Magnetism	Crystal structure
$MnFe_2O_4$	5,200	300	Ferrimagnetism	Spinel
$FeFe_2O_4$	6,000	585	Ferrimagnetism	Spinel
$NiFe_2O_4$	3,400	590	Ferrimagnetism	Spinel
$CoFe_2O_4$	5,000	520	Ferrimagnetism	Spinel
$CuFe_2O_4$	1,700	455	Ferrimagnetism	Spinel
$BaFe_{12}O_{19}$	4,000	450	Ferrimagnetism	Hexagonal
$SrFe_{12}O_{19}$	4,000	453	Ferrimagnetism	Hexagonal
$MgFe_2O_4$	1,400	440	Ferrimagnetism	Spinel
$Li_{0.5}Fe_{2.5}O_4$	3,900	670	Ferrimagnetism	Spinel
$\gamma\text{-}Fe_2O_3$	5,200	575	Ferrimagnetism	Spinel
$Y_3Fe_5O_{12}$	1,700	287	Ferrimagnetism	Garnet
$ZnFe_2O_4$			Antiferromagnetism	Spinel

feature of forming solid solutions with other spinel ferrites and causing the saturation magnetization to increase. Thus, we select the composition of these solid solutions according to our goal and application to produce a variety of useful ferrites. The major ferrite solid solutions are Mn–Zn ferrites ($Mn_xZn_{1-x}Fe_2O_4$), Ni–Zn ferrites ($Ni_xZn_{1-x}Fe_2O_4$), and Cu–Zn ferrites ($Cu_xZn_{1-x}Fe_2O_4$). These, along with the Ba and Sr ferrites, form the mainstream of magnetic materials for electronics applications.

III. SUMMARY OF CHARACTERISTICS

A. MAGNETIC CHARACTERISTICS

One of the basic properties characterizing ferro-magnetic materials is that expressing their magnetization process, that is, their hysteresis magnetization curves. Modeling the magnetization curve and extracting its distinctive features makes it possible to gain a conceptual understanding of these properties.

Figure 7.1 shows the relationship between the magnetic field strength H and the magnetic flux density B. The process, as the arrows indicate, is extremely complex. The form and size of such magnetization curves first vary depending upon the type and composition of the ferrite and then are also influenced by the mechanism of magnetization, that is, by the ways in which the initial, discontinuous, and rotation magnetization regions contribute to the shape of the hysteresis curve. The magnetic properties to be used for a particular application are closely related to the portion of this magnetization curve we are trying to use. Thus, it is highly important to

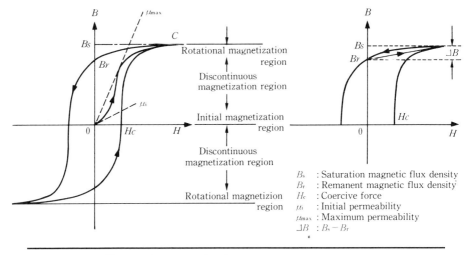

Figure 7.1. Hysteresis magnetization curves.

obtain a magnetization curve in line with the intended application. Therefore, selection of the composition of materials and the design of conditions to yield an ideal sintered body and to achieve microstructure control must also be suited to the application in question.

In practically useful materials, the choice of magnetic characteristics makes it possible to link the materials to a wide variety of applications. Qualitatively, there are materials which are easy to magnetize; these soft magnetic materials have a large permeability $\mu = B/H$. Materials with a large Br, for which considerable magnetization remains even when the magnetic field $H = 0$, are hard magnetic materials. And in between those two classes of materials are the semihard magnetic materials.

B. MICROWAVE CHARACTERISTICS

Some kinds of ferrites, in addition to having the magnetic characteristics described above, display other special phenomena due to the interaction of the microwaves and electron spin within the magnetic material. That is, if a microwave electromagnetic field is applied to ferrite in a static magnetic field, precession movement of the electrons is generated, and they slip from their natural spin directions. As a result, apart from the magnetization at less than several hundred megahertz described above, a completely new characteristic is added due to dynamic magnetization. This effect is the source of magnetic functions in microwave regions (Konishi, Kobayashi, and Tsuji, 1972).

This characteristic is due to the so-called gyromagnetic phenomenon of the electron spin direction in the ferrite and the high-frequency, circularly polarized magnetic field spinning in the same direction. The cause lies in a

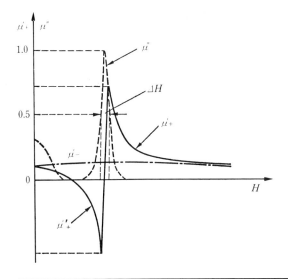

Figure 7.2. Circular polarization permeability and static magnetic field. From Konishi, Kobayashi, and Tsuji (1972, p. 168).

circular polarization permeability μ_+ (in the same spin direction) and μ_- (in the reverse spin direction) which show a large difference at or near the half-width of the magnetic resonance. The relationship between circular polarization permeability and the static magnetic field strength applied is shown in Figure 7.2. Therefore, the magnetic waves propagated through ferrite receive irreversible effects, including that of producing this large difference in pass resistance.

IV. SOFT MAGNETIC FERRITE

Soft magnetic ferrite is used for core materials for coils and transformers. Magnetic characteristics in a weak magnetic field depend, in general, on the shape and area of the hysteresis curve in the region of the origin point of the initial magnetization curve O → C. Ferrite, however, is usually high in specific resistance. Thus, magnetic waves penetrate it well and the dependent relationship is preserved up to the high-frequency band (called a Rayleigh loop and approximating in numerical formula an ellipse). As a result, ferrite has a constant permeability μ_i over a wide frequency band. There is, however, an upper limit to these frequencies. The limiting frequency f_0 is obtained, approximately, by the following relationship:

$f_0 \approx 1{,}000/\mu_i$ MHz

In the region of f_0, the permeability declines suddenly and the magnetic loss increases swiftly. This magnetic loss includes the rapid increase of residual loss in addition to eddy-current loss.

These characteristics are derived from the basic physical properties of ferrite; thus, magnetic cores (coils, transformers, for instance), should be used below f_0.

High permeability and low loss are desirable in a magnetic core. First, to illustrate the primary factors in obtaining high permeability, let us discuss the case of Mn-Zn ferrite, a representative example.

The basic makeup to obtain high permeability—other characteristics such as anisotropy and magnetostrictive characteristics are considered and selected for—are the conditions for a homogeneous, highly dense sintered body. Thus, not only must the materials used be tested on the assumption of high purity but also a special sintering method is applied to obtain the ideal sintered body.

An example is shown in Figure 7.3 (Araki and Okutani, 1979). The firing is performed so that the stoichiometric equilibrium conditions are fulfilled throughout the entire process of heating, reaction, and cooling. Fine points in this process include a temperature program to prevent the

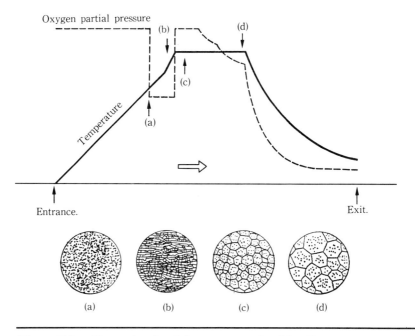

Figure 7.3. Example of a firing program for Mn-Zn ferrite and crystal growth. From Araki and Okutani (1979, p. 24).

minute amounts of impurities in the material to be sintered from being deposited in the grain boundaries (which would be the cause of anisotropy and hinder magnetization). In addition, to increase the fired density, the atmosphere is controlled during heating.

To minimize magnetic core loss, by contrast, requires different conditions from those for obtaining high permeability. A typical example of such materials is the reverse of the high permeability material discussed above: By forcing the ions of the impurities to be deposited in the grain boundaries, a thin layer with high resistance is formed, thereby yielding a sintered body with an overall high resistance. Desirable additives are silicon dioxide, calcium oxide (Akashi, 1961, p. 708), and titanium dioxide. To bring out their effects to the fullest, the establishment of a rational firing program is essential.

Figure 7.4 presents the distribution of calcium and silicon ions near the grain boundaries of ferrite obtained in this way Tsunekawa et al., 1979, p. 1857). The calcium ions with their large ionic radii show a striking segregation.

That these grain-boundary layers are high-resistance layers is verified by a separate method. They function to heighten the specific resistance of the sintered body as a whole. As a result, the relative loss coefficient tan δ/μ_i for the magnetic core shows remarkable improvement.

The permeability and magnetic core losses discussed thus far concern magnetic characteristics in extremely weak magnetic fields. Recently, however, applications have been rapidly opening up for which the magnetic characteristics near the saturation characteristics of the magnetization

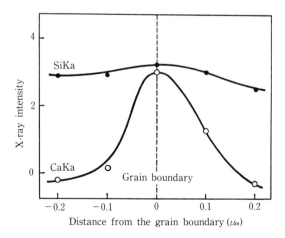

Figure 7.4. X-ray analysis of grain boundaries. From Tsunekawa et al. (1979, p. 1857). Copyright © 1979 IEEE.

curve are at issue. A fresh eye has been cast upon the characteristics of ferrite, due to the following circumstances.

Recently, with the goal of improving the efficiency of power supplies and making distributed processing practicable, there has been a swiftly growing trend of raising the frequencies of power supplies. (For instance, the frequency is raised from 50 Hz to 25 kHz for processing). The typical example is the switching regulator method.

The characteristic of ferrite sought there is its ΔB, which affects the loop near the saturation characteristic on the magnetization curve. Figure 7.1 partially indicates this relationship. This ΔB is related to the pulse permeability of the pulses used in the switching and to the magnetic core loss. Desirable conditions for a material with a large ΔB are, first, that the saturated magnetic flux density Bs be large and then that the residual magnetic flux density Br be small. Figure 7.5 shows the interrelationships between power loss and ΔB (Nomura, Okutani, Kitagawa, and Ochiai, 1982). To enlarge ΔB, measures true to the fundamentals of sintering are sought, including careful selection of the composition, curbing discontinuous grain growth, and producing a homogeneous sintered body with uniform grains so that the residual magnetization is slight.

Figure 7.6 indicates the course of improvements in power loss (Hiraga, 1982, p. 477). Thus, it has been possible to provide a foundation for the miniaturization of power supplies, an area which has lagged behind in electronic devices. Furthermore, there are further demands for reducing the size and weight of power supplies. Already the applied frequencies have progressed in the direction of higher frequencies, from 25 to 50 to 100

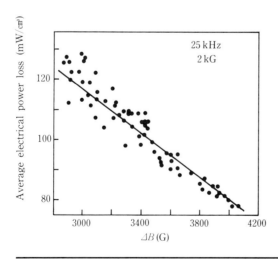

Figure 7.5. Relationship between average electrical power loss and ΔB.

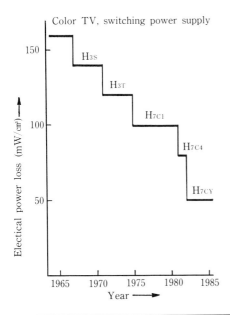

Figure 7.6. Trend in the reduction of electrical power loss for Mn–Zn ferrite (25 kHz, 2000 G). From Sōmiya (1982, p. 483).

to 300 kHz, and there is strong demand for further development of suitable materials.

Ceramics are hard and chemically stable. Compared with metallic magnetic materials, ferrite is superior not only in high-frequency magnetic characteristics but also in its additional special characteristics.

The classic applications of ferrite are in the magnetic heads of videotape recorders and computers. In those applications, not only is dimensional accuracy required but also there is a strong demand for both wear and environmental resistance. Metallic magnetic materials cannot fully meet these demands. As a result, ferrite has come to play the leading role in these fields.

Recently, however, the trend has changed from demand for high-density recording materials to demand for ferrite materials which can withstand high-precision processing. Even in the case of polycrystalline substances, it has already become necessary to make skillful use of high-level methods such as hot pressing or hot isostatic pressing instead of ordinary densification methods (Takama, 1982).

Manganese–zinc ferrites are used in the playback heads of videotape recorders. Given the necessity of making ultraprecision performance parts, which match up magnetic pieces of 20 μm thick to form, for instance, magnetic circuits with gaps only 0.35 μm long, single-crystal materials are used at present.

Such single crystals are usually cut from large-scale single crystals obtained by the Bridgman technique (Torii, Kihara, and Maeda, 1980, p. 717). Production techniques which incorporate measures to prevent compositional fluctuations are well established. Nonetheless, the process uses highly expensive platinum crucibles, and at times microscopic amounts of platinum are mixed into the crystals. Thus, it is not without its problems. A method of growing single crystals at a comparatively low temperature without using a crucible has been proposed to correct that point (Matsuzawa, 1982). Interest is focusing on its potential as an inexpensive method of forming single crystals.

V. HARD FERRITE

In the case of magnet materials (hard ferrites), a strong magnetization remains after a magnetizing field has been removed, and residual magnetization is stable even if a certain strength of demagnetizing field is applied. These are the characteristics that fit the image of the term *hard magnetic*. Materials with these characteristics are valuable in making permanent magnets.

In terms of the hysteresis curve in Figure 7.1, the first quadrant shows the magnetization process. For the applications of hard magnetic materials, greater significance lies in the demagnetization process in the second quadrant, that is, in the form of the demagnetization curve and how large its area is. For magnet materials, however, there is a considerable difference between the $B-H$ curve and the $I-H$ curve, in which the figure I shows the magnetization. The $I-H$ curve is important in magnet design and the evaluation of hard magnetic materials. Thus, it is necessary to distinguish these values accurately in handling them.

The quality of a magnet material depends upon the value of the magnetic energy. It is expressed as the product of the flux density B and the demagnetization field H at the operating point on the demagnetization curve. The maximum energy product $(B \times H)_{max}$ is usually used as a guideline for evaluation of materials for magnets.

The most important of these permanent magnet materials in practical use are barium ferrite ($BaO\text{-}6Fe_2O_3$) and strontium ferrite ($SrO\text{-}6Fe_2O_3$). Since they have a larger coercive force than do metallic magnetic materials, it is possible to design very thin magnets made of them. In addition, they have the distinctive features of low specific weight and favorable use of resources (since they do not require nickel, cobalt, etc.).

Compared with soft magnetic ferrite, hard magnetic ferrite is weak in structural sensitivity, and is relatively little influenced by impurities and by firing conditions. To obtain high coercive force with high density, however, it is essential for sintering to proceed at low temperatures, controlling grain growth. At times, additives such as silicon oxide or bismuth oxide must be used (van den Broek and Stuijts, 1977, p. 157).

Basically, however, the production method is not very different from that used for soft magnetic ferrite. Some specialized strategies are used, however, such as pressing in a magnetic field in advance for anisotropic magnets. At present the wet method (wet pressing in a magnetic field) has become the main force in improving the characteristics of anisotropic magnets.

Recent examples of materials development being sought include the value of the maximum energy product, but also there are strong requirements placed on the shape of the demagnetization curve. That is, the requirement is for a characteristic in which the second quadrant is expressed by a straight line, with Br and Hc each preferred at the 4,000 level. This requirement arises from the required stable operation of rotating machines such as motors without degradation.

To obtain this category of materials, the first requisite is careful choice of the raw materials and of the primary particles and finally the control of the microstructure of the sintered body.

Figure 7.7 indicates the progress, organized by year, in developing this series of hard magnetic ferrites. It also includes developments predicted for the near future.

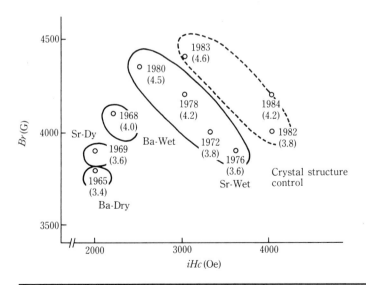

Figure 7.7. *Progress in the maximum energy product for anisotropic ferrite magnets. Numbers next to data points (open circles) show years, followed by the energy product MGOe in parentheses. From Sōmiya (1982, p. 485).*

In the search for improved characteristics, research on W-type barium ferrite (BaFe^{2+}Fe$^{3+}_{16}$O$_{27}$) has made progress (Lotgering, Vromans, and Huyberts, 1980, p. 5913). Its existence was known, but little attention was paid to it because of the difficulty of stabilizing the W phase when fired in an air atmosphere. But its saturation magnetization is 10% higher than that of the conventional M type, it offers potential for increasingly high performance, and there has been progress in the technology for atmospheric control in firing. These factors have led to a reconsideration of the material. Magnetic characteristics of Br = 4,000 to 4,700 and $(B \times H)_{max}$ = 3.7 to 4.3 MGOe have already been obtained. Whether it will reach a position as a useful magnetic material, especially in terms of costs, however, is still an open question.

VI. SEMIHARD MAGNETIC FERRITE

Semihard magnetic ferrite, which is used for memory and magnetic recording applications, falls between hard and soft magnetic ferrite. Its distinctive features are that it is comparatively easy to magnetize, can retain magnetization, and can produce output (changes in the magnetic flux) from this state, as needed.

The typical application is in memory cores for computers. An ultrasmall toroidal magnetic core is used in memory cores. The two remanent magnetization states $+Br$ and $-Br$ are set equivalent to 1 and 0 and are made to function as the smallest memory units. Their significant characteristic is the magnetic reversal time t_s, that is, higher speeds for $+Br \rightleftarrows -Br$ are desirable. Therefore, a material with a good rectangularity Br/Bm is needed. In addition, for practical purposes, t_s has the following relationship with the hysteresis curve:

$$t_s \approx K_s/kHc$$

where K_s is called the switching coefficient. It is determined by the materials in the ferrite, the grain size, and so on.

In general the Mn–Mg ferrites as well as the lithium ferrites, which are outstanding in speed and temperature characteristics, are used in memory cores. They have become increasingly prominent in applications requiring highly reliable memory with the special feature of nonvolatility. (The memory is not erased even when the power supply is down.) In recent years, however, the amazing progress in semiconductors has led to great changes in leadership in memory elements.

Magnetic recording also puts to use the same types of characteristics needed in memory cores. The magnetic materials used in magnetic recording tape are different from the sintered bodies described thus far, because the powder itself is the material with the desired features for the application.

That is, for recording applications, the necessary conditions for this magnetic powder are that it must be stable, have a high coercive force, and be a fine powder suitable for making into a paint. There were, however, limits on the increases in coercive force obtained by simple pulverization methods. In addition, the stability of the coervice force (temperature dependence, for instance) have also been a problem.

These problems were solved by M. Camuras' discovery of acicular or needle-form γ-Fe_2O_3 (Camuras, 1951). A brief description of this substance follows.

Like Fe_3O_4, γ-Fe_2O_3 belongs to the spinel crystal group. Therefore, in terms of coercive force, it had some problems with a small crystal anisotropy. When made into a long, fine, acicular form, however, it was possible to add shape anisotropy to it. The problem was how to succeed in producing acicular γ-Fe_2O_3. The method is special. It begins by using α-FeOOH, a mother salt with an excellent ratio of needle-shaped crystals. In the acicular form of γ-Fe_2O_3, that of the mother salt is preserved as framework particles. The resulting substance must be able to withstand the process of being made into paint and into tape (with the needles aligned in the long direction). Normally the acicular form ratio appears to be on the order of $8:1$.

In the last few years, however, as the technical trend in magnetic recording has been from audio to video and in the direction of increasingly higher recording densities, signs have been appearing that there may be a limit to the densities attainable by using acicular γ-Fe_2O_3. Acicular CrO_2 has been proposed for such applications. Even though it is a metal oxide material outside the ferrite family, its significance requires special mention. To be brief, γ-Fe_2O_3 was replaced by a composite material; that change is directly tied to the strength of Japan's magnetic tape industry. That composite material is called cobalt-adsorbed γ-Fe_2O_3.

The new magnetic powder uses γ-Fe_2O_3, with its excellent acicular form ratio, as the nucleus upon the surface of which a cobalt content layer tens of angstroms thick is formed. The submicrometer order double-layer structure means that the cobalt is not allowed to diffuse within the nucleus and form a solution with it. This structure is the key to manifesting the material's special features. See Figure 7.8 for an analysis of the structure (Tokuoka, Sugihara, Oka, Imaoka, 1981, p. 1564). The cobalt-bearing layer extremely near the particle surface contributes to the induction of surface anisotropy. In addition, at a cobalt content of only about 3%, the coercive force is saturated, which indicates the great significance of the cobalt's location on the surface. The mechanism for this induction of surface anisotropy, however, is not yet fully understood.

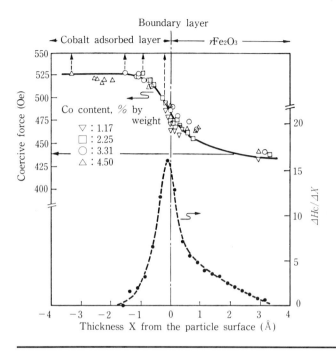

Figure 7.8. Cobalt adsorbed layer thickness and coercive force. From Tokuoka et al. (1981, p. 1564).

In recent years the basic facts of magnetic recording have been reconsidered. Research and development projects from the perspective that a vertical magnetic recording method is suited to higher density recording are now in process. (Iwasaki and Ouchi, 1978, p. 849). The Co–Cr sputter film is one such recording medium that has been developed.

To return to the discussion of Ba ferrite, it belongs to the hexagonal system and exists as single particles which are thin hexagonal plates with an easily magnetized axis in their thickness direction. Therefore, if these fine particles can be made into a thin layer on a plane, they are likely to produce a medium with constituents easily magnetized vertically. Currently, magnetic cards, for instance, use a barium ferrite powder with high coercive force (Hosaka, Tochihara, and Namikawa, 1980, p. 575); they have achieved high reliability, so that the recorded data do not disappear. This application, however, was not really made with an eye to the vertically magnetized constituent.

Recently, however, a new medium proposed for vertical recording that would use ultrafine particles of barium ferrite in a special technique has been attracting attention (Kubo, Ido, and Yokoyama, 1982). This barium ferrite is made using the crystallization of glass method to obtain a medium with plate-form particles 0.08 μm in diameter with a coercive force of 900 Oe and a high saturation magnetization. Experimental objects using it do not equal the recording characteristics of Co–Cr vertically magnetized film (made using the sputter method), but they equal the performance of ordinary metal tape. They offer great promise as a new recording method by stable oxides.

VII. FERRITE FOR MICROWAVE USE

The materials discussed thus far can be regarded as having been developed by careful selection of the magnetization characteristics which affect their hysteresis curves. Microwave applications, however, use the gyromagnetic phenomenon mentioned above; their situation is somewhat different. The basis of the phenomenon is the active use of the magnetic saturation region of ferrite, though, so that there is no great difference in the thinking about the materials or about their functions from what has come before.

The essence of the gyromagnetic phenomenon is presented in Figure 7.2. The decrease in the ferro-magnetic linewidth (ΔH), which indicates how good the magnetic resonance is, and the difference between μ'_+ and μ'_-, the circular polarization permeability, are the most conspicuous ways of evaluating this characteristic. In measures to reduce ΔH, it is critical to exclude the major factors which prevent magnetization, that is, crystal anisotropy, porosity, and nonmagnetic impurities.

Also, in general, the magnetic resonance frequency will be proportional to the applied static magnetic field. Thus, to draw out the functional characteristics of this phenomenon effectively, it is necessary to prepare a material with different saturation magnetization, in order to almost saturate it with this applied static field. This condition implies that it is quite important to control the saturation magnetization of the material.

The materials used include those in spinel structure, hexagonal system, and garnet structure; each is used in different applications. Here we describe only some of their characteristics.

Figure 7.9 indicates the relationship between the fired density of garnet-structure polycrystalline yttrium iron garnet and ΔH (Rodrigue *et al.*, 1958, p. 83). It points out the importance, first of all, of producing a high-density sintered body.

Figure 7.10 shows the great change in saturation magnetization caused by replacing part of the Fe with Al in Mg–Mn ferrite (van Uitert, 1955, p. 1289). Thi replacement does, however, have the practical problem of leading to an excessive drop in the Curie point. Other constituents are being considered. The technical trends presented in the next figure are pertinent to this issue.

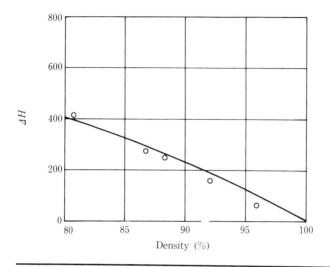

Figure 7.9. Relationship between ΔH and the density of polycrystalline yttrium iron garnet.

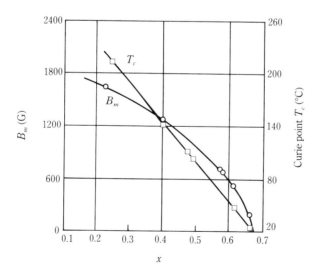

Figure 7.10. Saturation magnetization flux density and Curie point for Al-substituted Mg–Mn ferrite (for $MgMn_{0.1}Al_xFe_{2-x}O_4$).

The classic devices using microwave characteristics are irreversible, typically microwave circulators and isolators. In a circulator, three terminal elements are placed at 120° distances. The electromagnetic waves pass in the direction from terminal one to two to three, but not in the reverse direction. In an isolator, as well, the waves pass in the forward direction but not in reverse. Thus, these devices exhibit the irreversible function. Related types include the magnetic resonance isolator, which uses the magnetic resonance phenomenon.

Figure 7.11 summarizes technical trends in materials for microwave applications in general from the point of view of applicable frequencies and low loss (Hiraga, 1976, p. 970). The technological background in each decade is clearly reflected in the types of materials developed and the characteristic values for each.

Recently, the development of microwave technology has made it a part of our ordinary lives. Examples include the marketing of automobile telephones, which are the typical example of mobile radio, the possibility of nationwide television broadcasting and high-definition television in Japan through satellite broadcasts, and the widespread use of sensors

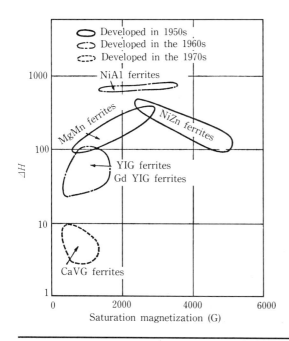

Figure 7.11. Development trend for ferrite used in microwave applications. From Hiraga (1976, p. 971).

which employ microwave characteristics. Trends for future development include the integration of microwave devices.

This chapter has focused on ferrite, the representative oxide magnetic material. The discussion has largely applied to sintered materials. Magnetostrictive phenomena and special thin-film effects (such as bubble memory) have not been touched on.

Ferrite materials perform from low to microwave frequencies, as has been explained. Naturally, this extension reaches the region of light waves, and ferrite (principally the garnet structure) is being tested as a promising material there as well.

In a rather different field, magnetite has been formed in some types of bacteria, and its connections with biotechnological properties are arousing interest. The points of contact between magnetism and light and magnetism and biology are fields that offer great promise for development.

Several excellent specialized books on ferrite have already have published. Thus, the interested reader is urged to refer to them for a fuller explication of this subject.

REFERENCES

Akashi, Masao. (1961). Mn-Zn feraito kesshō ryūkaisō no denki teikō [Electric resistance at the grain boundaries in Mn-Zn ferrite crystals.] *Journal of Applied Physics in Japan,* 30, 708–709.

Araki, Shigeo and Okutani, Katsunobu. (1979). Feraito (2)—Feraito Kōgyō no genjō [Ferrite (2)—the state of the ferrite industry.] *Nippon oyo jiki gakkai,* 3, (1), 20–26.

Camuras, M. (1951). Japanese patent Shōwa 26-776.

Denshi zairyō kōgyōkai (ed.). (1972). *Feraito no ōyō gijutsu* [Applications technology for ferrite]. Tokyo: Seibunsha.

Gorter, E. W. (1955). Some properties of Ferrites in Connection with their Chemistry. *Proceedings of the I. R. E.* 43, 1945–1973.

Hiraga, Teitarō. Feraito (buhin zairyō no 20 nenkan no shinpo, hatten) [Ferrite: 20 years of progress and development as a component material] *Denshi tsūshin gakkai shi,* 59, 970.

Hiraga, Teitarō. (1982). Feraito no hatsumei to sono hatten (Discovery and development of ferrite). In Sōmiya, Shigeyuki (ed.), *Seramikkusu no kagaku to gijutsu no genjo to shorai.* Tokyo: Uchida-Rokakuho, p 477.

Hosaka, Hiroshi; Tochihara, Shigezō; and Namikawa Mamoru. (1980). Digital recording properties of high coercivity magnetic credit card. In *Proceedings of the 3rd International Conference on Ferrites,* 575.

Iwasaki, Shun'ichi and Ouchi Kazuhiro. (1978). Co–Cr recording films with perpendicular magnetic anisotropy. *IEEE Transactions Magazine,* Mag-14, 849–851.

Katō, Yogorō and Takei, Takeshi. (1932). Japanese patent 98,844.
Konishi, Yoshirō; Kobayashi, Seihin; and Tsuji, Toshirō. (1972). *Feraito erekutoronikkusu* [Ferrite electronics]. Tokyo: Nikken kogyo shinbunsha.
Kubo, O.; Ido, T.; and Yokoyama, Hiroshi (1982). Properties of Ba ferrite particles for perpendicular magnetic recording media. Paper presented at the Intermag MMM Conference, July.
Lotgering, F. K.; Vromans, P. H. G .M.; and Huyberts, M. A. H. (1980). Permanent-magnet material obtained by sintering the hexagonal ferrite. *Journal of Applied Physics,* 51, 5913–5918.
Matsuzawa, Motoichirō. (1982). Mn–Zn ferraito no kosō ka ni okeru tan kesshō ikusei [Growing single crystals under solid phase Mn–Zn ferrite]. Paper presented at Daiikkai denshi zairyo toronkai, Yogyo kyokai, Autumn.
Mitō kakō gijutsu kyōkai (ed.). (1981). *Shinjidai no jisei zairyo* [Magnetic materials for a new age]. Tokyo: Kōgyō chōsakai.
Ne'el, L. (1948). *Magnetic Properties of Ferrites: Ferrimagnetism and Antiferro magnetism. Annales de Physique,* 3, (12), 137–198.
Nomura, Takeshi; Okutani, Katsunobu; Kitagawa, Takeo; and Ochiai, Tatsushiro. Sintering of Mn–Zn ferrites for power electronics. Paper presented at the Fall Meeting of The American Ceramics Society, Boston.
Rodrigue, G. P., et al. (1958). Ferrimagnetic resonance in some polycrystalline rare earth garnets. *IRE Transactions, MTT-6,* 83–91.
Snoek, J L. *New developments in ferromagnetic materials.* (1947). New York: Elsevier.
Sōmiya, Shigeyuki (ed.). (1982). Seramikkusu no Kagaku to gijutsu no genjo to shorai [The present state and future for the science and technology of ceramics]. Tokyo: Uchida Rokakuho.
Takama, Eizō. (1982). Nekkan seisuiatsu puresu ni yoru kōmitsudo mangan aen feraito [High density manganese zinc ferrite achieved by hot hydrostatic pressing]. *Nippon ōyō jiki gakkai,* 6, (3), 185–187.
Takei, Takeshi. (1960). *Feraito no riron to ōyō* [Theory and applications of ferrite]. Tokyo: Maruzen.
Tokuoka, Yasumichi; Sugihara, Yō; Oka, Yukihiro; and Imaoka, Yasuo. (1981). Kobaruto kyūchaku α-Fe_2O_3 ryushi hyomen kozo no netsu shori ni yoru henka [Changes due to heat treatment in the surface of grains of cobalt-adsorbed α-Fe_2O_3]. *Nikkashi,* 10, 1564–1570.
Torii, Michiro; Kihara, Utsuo; and Maeda, Ikuo. (1980). New process to make huge spinel ferrite single crystals. In *Proceedings of the 3rd International Conference on Ferrites,* 1873–1875.
Tsunekawa, Hiroshi, et al. (1979). Microstructure and properties of commercial grade manganese zinc ferrites. *IEEE Transactions Magazine, Mag-15,* 1855–1857.
van den Broek, C. A. N., and Stuijts, A. L. (1977). Ferroxdure. *Philips Technical Review,* 37, (7), 157–175.
van Uitert, L. G. (1955). Low magnetic saturation ferrites for microwave applications. *Journal of Applied Physics,* 26, 1289–1290.

8
Thermal Properties

Tatsuyuki Kawakubo
Tokyo Institute of Technology
Meguro-ku, Tokyo 152, Japan

Noboru Yamamoto
NGK Insulators Co., Ltd.
Mizuho-ku, Nagoya 467, Japan

I. THEORY OF THERMAL PROPERTIES

In general, the basic thermal properties of all matter, including ceramics, are specific heat, thermal expansion, and thermal conductivity. If thermal properties are considered in relation to applications from an engineering standpoint, other phenomena associated with heat and electricity become significant, including thermoelectric effect, pyroelectricity, and thermionic radiation. Here, as a preliminary step, the discussion focuses, however, on only the phenomenon of heat itself and the three basic properties.

A. SPECIFIC HEAT

When the specific heat of a solid is measured, the measurements are usually carried out at a certain presssure; thus the measured value is the constant-pressure specific heat C_p. The specific heat calculated theoretically is, however, the constant-volume specific heat C_V. The difference

Note: Tatsuyuki Kawakubo wrote section I; section II is by Noboru Yamamoto.

between these two specific heats is given by Eq. (8.1)

$$C_P - C_V = \frac{\alpha^2 V_0 T}{\chi} \tag{8.1}$$

Where α is the volume thermal expansion coefficient, V_0 the molar volume, T the absolute temperature, and χ is the compressibility, where $\chi = -dV/Vdp$. According to the first law of thermodynamics, if the heat energy supplied externally is dQ, the increase in internal energy in the matter due to it is dU, the thermal expansion is dV, and the external pressure is p, then, since Eq. (8.2) obtains,

$$dQ = dU + p\,dV \tag{8.2}$$

the constant-volume specific heat is given by the changes in the internal energy due to temperature, as in Eq. (8.3).

$$C_V = \left(\frac{\partial U}{\partial T}\right)_V \tag{8.3}$$

In contrast, for the constant-pressure specific heat, the amount of increase in the internal energy through work carried out outside and through thermal expansion must be added to the value calculated with Eq. (8.3). Thus, as is shown in Eq. (8.1), there is a difference between constant-pressure specific heat and constant-volume specific heat. Since this difference, as a percentage of the constant-volume specific heat, is of the order of 0.6% for tungsten and 2.8% for copper, we should first consider the constant-volume specific heat as expressed by Eq. (8.3).

Each atom in a solid thermally oscillates around its equilibrium point. These oscillations are called lattice vibrations. Lattice vibrations are the major part of the solid's internal energy that changes according to temperature. In solids, particularly in metals, there are free electrons matching the atoms in number. Their kinetic energy makes a great contribution to the solid's internal energy, but since the electrons follow the Fermi–Dirac statistics, that internal energy changes very little in response to changes in temperature. Thus, the specific heat of ordinary solids can be regarded as being due to the lattice vibrations alone.

When the atoms are in harmonic oscillation around their equilibrium points, according to the law of equipartition of energy, for one degree of freedom the average kinetic energy and the average potential energy will both be $(\frac{1}{2})k_B T$. Thus, if the number of atoms is N and the oscillation is three-dimensional, the energy of the whole body will be given by Eq. (8.4):

$$U = 3Nk_B T \tag{8.4}$$

Therefore, if Avogadro's number is denoted by N_A, the constant-volume specific heat is found with Eq. (8.5).

$$C_V = \left(\frac{\partial U}{\partial T}\right) = 3N_A k_B = 3R \approx 24.9 \text{ J/deg·mol} \tag{8.5}$$

That is called Dulong–Petit's law. In the region of room temperatures, the specific heat of monatomic solids does indeed have this value. When the temperature drops, however, the specific heat declines; near absolute zero the specific heats of all solids are known to vary as the cube of the absolute temperature.

Einstein was the first person to attempt a theoretical explanation of this deviation of the specific heat from what is predicted by Dulong–Petit's law and its decrease. Since Debye then refined his explanation, we consider Debye's theory of specific heat here. First, in classical mechanics, the energy of a harmonic oscillator with the angular frequency ω can take on any value in a continuum no matter how finely subdivided, according to differences in the amplitude. Quantum theory states, however, that only discrete values can occur, as given in Eq. (8.6).

$$E_n = (n + \tfrac{1}{2})\hbar\omega \quad (n = 0, 1, 2 \ldots) \tag{8.6}$$

In this equation, h is Planck's constant. In this case the probability that an oscillator will have any given one n of the values given in Eq. (8.6) is proportional to $\exp[-(n + \tfrac{1}{2})\hbar\omega/k_B T]$. Thus, the average energy of a single oscillator is found with Eq. (8.7).

$$\langle E \rangle_{AV} = \frac{\sum_{n=0}^{\infty}(n + 1/2)\hbar\omega \exp[-(n + 1/2)\hbar\omega/k_B T]}{\sum_{n=0}^{\infty}\exp[-(n + 1/2)\hbar\omega/k_B T]} \tag{8.7}$$

$$= \frac{1}{2}\hbar\omega + \frac{\hbar\omega}{\exp(\hbar\omega/k_B T) - 1}$$

Thus far we have assumed that all oscillators would be vibrating at the same frequency ω. Actually, however, oscillators with many different frequencies and wavelengths exist. If the number of oscillators with frequencies between ω and $\omega + d\omega$ is $g(\omega)\,d\omega$, to find the internal energy U, we do not simply multiply the average energy found in Eq. (8.7) by $3N_A$. Rather, we must multiply by the weighting factor $g(\omega)$ and integrate for ω. The $g(\omega)$ of elastic waves in solids is proportional to ω^2; Debye argued that lattice vibrations in solids approximate elastic waves and that, as a corollary of the fact that there are no oscillations with wavelengths shorter than the interatomic distances, there are no oscillators with frequencies greater than some maximum value ω_m. Thus, he proposed the weighting factors in Eq. (8.8).

$$g(\omega) = \begin{cases} \dfrac{9N_A}{\omega_m^3}\omega^2 & \omega < \omega_m \\ 0 & \omega > \omega_m \end{cases} \tag{8.8}$$

The coefficient $9N/\omega_m^3$ is a normalized constant for $g(\omega)$ set to have the value of $3N_A$, the number of kinetic degrees of freedom, when $g(\omega)$ is integrated over 0 to ω_m. Using it, we can find the internal energy with Eq. (8.9).

$$U = \frac{9N_A}{\omega_m^3} \int_0^{\omega_m} h\omega^3 \left\{ \frac{1}{2} + \frac{1}{\exp(h\omega/k_B T) - 1} \right\} d\omega \tag{8.9}$$

Then, differentiating for T, we are given the specific heat in the form given in Eq. (8.10).

$$C_V = \frac{9N_A}{\omega_m^3} \int_0^{\omega_m} \frac{h^2\omega^4 \exp(h\omega/k_B T)}{\{\exp(h\omega/k_B T) - 1\}^2 k_B T^2} d\omega \tag{8.10}$$

Let $h\omega/k_B T = x$. Then using Debye's specific temperature, Θ_D, defined as

$$h\omega_m/k_B \equiv \Theta_D \tag{8.11}$$

Eq. (8.10) can be rewritten as Eq. (8.12).

$$C_V = 9N_A k_B \left(\frac{T}{\Theta_D}\right)^3 \int_0^{\Theta_D/T} \frac{e^x x^4}{(e^x - 1)^2} dx \tag{8.12}$$

Calculating numerical values from it produces the curve given in Figure 8.1. It fits experimental results extremely well.

B. THERMAL EXPANSION

Setting aside phase transformations, under normal conditions, all substances undergo an increase in volume with increased temperature. The cause, to state it briefly, arises from the potential anharmonic character of the oscillation of the atoms which make up the substance. Taking R_0, the distance between atoms obtained at absolute zero, as the standard, the increases in potential energy when those atomic distances are expanded

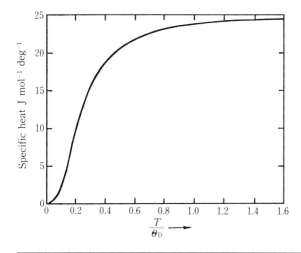

Figure 8.1. Specific heat curve according to Debye's model.

and when they are contracted are, as Figure 8.2 shows, asymmetrical. The temperature rises, vibrations are excited, and the average value of the distance between atoms slips toward the larger direction, to give a qualitative explanation of thermal expansion. If r is the slip from the equilibrium value for the distance between atoms R_0, the increase in potential energy in

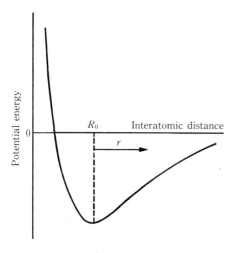

Figure 8.2. Potential energy vs. interatomic distance.

the region in which r is small approximates the value as given in Eq. (8.13).

$$V(r) = ar^2 - br^3 \tag{8.13}$$

In this case the average value for r at temperature T is given in Eq. (8.14).

$$\langle r \rangle = \frac{\int_{-\infty}^{\infty} r \exp\{-(ar^2 - br^3)/k_B T\} dr}{\int_{-\infty}^{\infty} \exp\{-(ar^2 - br^3)/k_B T\} dr} \tag{8.14}$$

$$= (3b/4a^2)k_B T$$

Due to the nonlinear term $-br^3$, it is deduced that the distance between atoms will increase with the temperature.

In the above model, expansion occurs in proportion to temperature, so that the coefficient of thermal expansion is a constant, not affected by temperature. Since the potential energy within actual solids includes a nonlinear term of higher degree, the coefficient of thermal expansion does exhibit dependency on temperature. If there is nonlinearity, the upper limit to the frequency of the lattice vibrations changes with expansion and therefore Debye's specific temperature Θ_D changes. Seeking the characteristic equation for a solid with this conclusion in mind leads to a proportional relationship between the thermal expansion coefficient α and the specific heat C_V,

$$\alpha = \frac{\gamma \chi C_V}{V} \tag{8.15}$$

for an isotropic crystal. Here γ is given by

$$\gamma = \frac{\partial \log \Theta_D}{\partial \log V} \tag{8.16}$$

It is a physical constant called the Grüneisen constant . Equation (8.15) works well for isotropic crystals. At the low temperatures at which the specific heat becomes smaller, α, paralleling it, also becomes smaller.

The expansion coefficient for anisotropic crystals varies depending upon the crystallographic axis. In general, as the temperature rises, the lattice constants change in the direction of less anisotropy. Particularly interesting are cases of crystals with a layer structure, such as graphite. In graphite, the linear expansion coefficient perpendicular to the c axis is overwhel-

8. Thermal Properties   151

mingly smaller than the coefficient for the c axis. In addition, in Al_2TiO_5, the linear expansion coefficient perpendicular to the c axis is negative; as a result, the coefficient of cubic expansion is extremely small.

C. THERMAL CONDUCTIVITY

The carriers of thermal conductivity in solids are the lattice vibration waves and the motion of the electrons. Excluding metals and metallic alloys, however, in electrical insulators and most semiconductors, the contribution of electrons to the conduction of heat is small. Lattice vibrations become waves and transmit the heat. The energy of lattice vibrations can, as Eq. (8.6) shows, take on only discrete values. Thus, when those become waves and are transmitted, the wave with wave number vector q is excited and quieted in energy quantum units $h\omega$ (q). That is, when the lattice vibration waves collide with each other and when lattice vibrations and electrons reciprocally exchange energy, the energy is in $h\omega$ (q) quantum units. The quanta of lattice vibration waves are called phonons. These are quite similar to photons, which are quantized electromagnetic waves.

Thus, the carriers for thermal conduction in insulators and semiconductors are phonons. Consider the behavior of phonons in a solid at a certain temperature. Phonons are constantly being born and disappearing, but their average number is fixed. The wave number q and, therefore, the number of phonons with energy $h\omega$ (q) are, based on a comparison of Eqs. (8.6) and (8.7), given by Eq. (8.17).

$$n_q = \frac{1}{\exp(h\omega(\mathbf{q})/k_B T) - 1} \qquad (8.17)$$

If no temperature gradient exists in the test material, this amount will be the same throughout the crystal, and no flow of heat will occur. If, however, there is a temperature gradient in the test material, there will be a difference in the energy distribution of the phonons according to their location within the crystal. To equalize this difference, the phonons will spread from a location of high density to one of low density. This phenomenon is the mechanism of thermal conductivity. While spreading, phonons will collide with other phonons, be scattered by impurities and lattice defects, and will move with a repeated zigzag motion. The average distance from one collision to the next is the mean free path, l. Using that variable, the average velocity of the phonons, v, and the specific heat C for a unit volume yields the thermal conductivity given in Eq. (8.18).

$$\kappa = \frac{1}{3} Cvl \qquad (8.18)$$

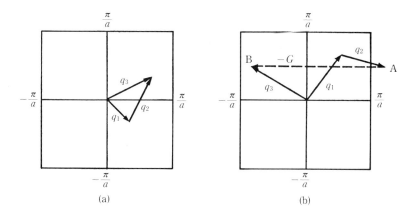

Figure 8.3. First Brillouin Zone in the two-dimensional square lattice with interatomic distance a and the phonon collision process. (a) shows the normal process $\mathbf{q}_1 + \mathbf{q}_2 = \mathbf{q}_3$. (b) shows the Umklapp process $\mathbf{q}_1 + \mathbf{q}_2 - \mathbf{G} = \mathbf{q}_3$; here, $\mathbf{G} = 2\pi/a$. In (b), the direction of $\mathbf{q}_1 + \mathbf{q}_2$ is opposite to that of $\mathbf{q}_3$. This process is the origin of thermal resistance.

This equation is derived for the thermal conductivity of a gaseous molecule, but phonons can be regarded as collections of particles, just as collections of molecules are, so that Eq. 8.18 can be applied as-is.

The length of the mean free path l is determined by the collisions of the phonons with each other, by the impurities, and, particularly in the case of ceramics, by the grain size. As Figure 8.3 (a) indicates, the phonons with wave number vector $\mathbf{q}_1$ and $\mathbf{q}_2$ collide, producing a phonon with wave number vector $\mathbf{q}_3$. Energy conservation before and after the collision is always preserved (Eq. (8.19)).

$$h\omega_1 + h\omega_2 = h\omega_3 \tag{8.19}$$

Equation (8.20) for the crystal momentum may or may not hold.[1]

$$h\mathbf{q}_1 + h\mathbf{q}_2 = h\mathbf{q}_3 \tag{8.20}$$

The cases for which it does apply are called the normal process. In this case, as indicated in Figure 8.3 (a), there is no change in the direction or magnitude of the energy flow. This type of collision does not lead to a

---

[1] $h_\mathbf{q}$ is called the crystal momentum. It has acquired this name from the correspondence of Eq (8.20) to the law of conservation of momentum, but $h_\mathbf{q}$ is not momentum in the usual sense of the term in mechanics.

decline in the heat flow. That is, in the normal process, there is no heat resistance and the thermal conductivity is infinitely large.

In contrast, Peierls pointed out the importance to the occurrence of heat resistance of the process which fulfills the conditions given in Eq. (8.21),

$$\mathbf{q}_1 + \mathbf{q}_2 = \mathbf{q}_3 + \mathbf{G} \tag{8.21}$$

in which the wave number vectors before and after the collision are linked by the reciprocal lattice vector $\mathbf{G}$. As Figure 8.3 (b) shows, the vector with negative sign, $-\mathbf{G}$, is the vector which draws the combined wave number vectors $\mathbf{q}_1 + \mathbf{q}_2$ back into the first Brillouin zone when they have been forced out of it.[2] As explained in the footnote, the spatial pattern of the lattice wave for wave number vectors $\mathbf{q}_1 + \mathbf{q}_2$ is virtually the same as the pattern for the lattice wave $\mathbf{q}_3 = \mathbf{q}_1 + \mathbf{q}_2 - \mathbf{G}$. As Figure 8.3 (b) indicates, however, the direction of energy flow is opposite for $\mathbf{q}_3$ and for $\mathbf{q}_1 + \mathbf{q}_2$. Peierls named this type of collision an Umklapp process.

Thus, among phonon-phonon collisions, only the Umklapp process collisions give a length l of the mean free path that participates in thermal resistance. In normal process collisions, observing one phonon yields zig-zagging movements, but if both parties to the collision are considered together, the same direction and magnitude of heat flow are preserved before and after the collision.

Then, let us consider whether the mean free path l in the Umklapp process varies with the temperature. The Umklapp process occurs when, as a result of a phonon-phonon collision, the wave number vector goes outside the first Brillouin zone and the other phonon in the collision has a wave number vector on the order of $\mathbf{G}/2$, that is, an energy of the order of $k_B \Theta_D/2$ (where Θ_D is the Debye temperature). Since the percentage of collisions is proportional to the number of phonons, the mean free path may be regarded as being inversely proportional to the number of

---

[2] If the unit vectors in the x and y directions are i and j, then $(2\pi/a)i$, $(2\pi/a)j$ gives the basic vectors for the reciprocal lattice for a two-dimensional tetragonal lattice in which the lattice spacings on the x and y axes are both a. The square in Figure 8.3 encompassed by the perpendicular bisectors of this basic vector is called the first Brillouin zone. Point A, which is outside the first Brillouin zone, is in the same state of atomic displacement as point B, which is within the first Brillouin zone and has diverged just $-\mathbf{G} = -(2\pi/a)i$ away from point A. Their displacements are the same because, in actual space, the displacement of the atoms at the lattice point (m, n) is, for the wave vector at point A,

$$U_A(t) = U \cos(ma q_{3x} + na q_{3y} - \omega t)$$

and is, for the wave vector at point B,

$$U_B(t) = U \cos\{ma(q_{3x} - 2\pi/a) + na q_{3y} - \omega t\}$$

Thus, the two values are indeed the same.

phonons. To find the number of phonons with energy on the order of $k_B\Theta_D/2$, $k_B\Theta D/2$ may be substituted for the value of $\hbar\omega(\mathbf{q})$ in Eq. (8.17). Thus, at higher temperatures for which $T \gg \Theta_D$, the mean free path is proportional to $\Theta_D/2T$. At lower temperatures for which $T \ll \Theta_D$, it is proportional to $\exp(\Theta_D/2T)$.

At even lower temperatures, the percentage of phonon-phonon collisions grows steadily smaller, and other effects will determine the mean free path. Those are the scattering of phonons by collisions with impurity atoms, with lattice defects, and with the boundaries of the crystal grains. These scattering processes are, by and large, not affected by temperature. Thus, at low temperatures in which scattering processes play the major role, the mean free path is saturated, and the overall temperature dependence of the mean free path becomes typically as shown in Figure 8.4(a).

The absolute value of l cannot be calculated a priori. Rather, using Eq. (8.18) from actual measured values of thermal conductivity, the value

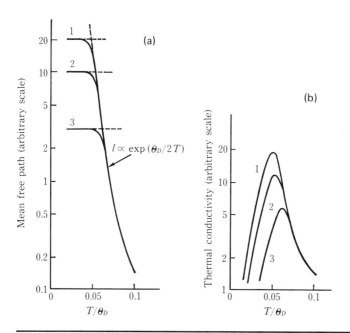

Figure 8.4. The mean free path of phonons (a) and thermal conductivity (b) in the low-temperature region. Curves 1, 2, and 3, in that order, indicate that the mean free path is reduced at low temperatures due to an increase in lattice defects or impurities.

of *l* can be evaluated. According to these results, in a single crystal of high purity, at room temperature *l* will be in the range of several to 10 or so lattice spacings in length. It is known, however, that at low temperatures the mean free path may reach a length of the order of 1 mm. In ceramics, the saturation value at low temperatures for the mean free path of phonons may be considered as determined by the crystal grain size. Thermal conductivity is the product of the mean free path, the specific heat C, and the phonon velocity v, as shown in Eq. (8.18). In theory v ought to be somewhat dependent upon temperature, but actually those changes are trivial. The specific heat, by contrast, is proportional to T^3 at low temperatures and stable at high temperatures, as described in section A. The temperature dependence of thermal conductivity, the product of C, v, and l, is given in Figure 8.4 (b). Thermal conductivity has its maximum value at low temperatures (in the region of 5–100 K). The maximum value is determined by the value l, the mean free path, reaches. That in turn is set by a number of factors discussed above, including lattice defects, impurities, crystal grain size, and test sample size. In the case of solid solutions, l is smaller than in substances of pure composition. In the high-temperature range, however, thermal conductivity is proportional to T^{-1}, and for the same substance the above lattice defects and other conditions of the production of the test specimen lose their influence.

This treatment of the temperature dependence of thermal conductivity largely explains the experimental data for most substances, but it is incapable of accounting for the absolute value of thermal conductivity. In particular, BeO is known to have a large thermal conductivity despite its lacking electrical conductivity, a phenomenon which is still a puzzle. That anomaly aside, in substances with high electrical conductivity which are close to the metals, such as carbides of transition metals, the electrons concerned in the electrical conductivity also contribute to thermal conductivity, so that electrical and thermal conductivity have a parallel relationship. Also, in substances in the oxide group, at high temperatures (1500°C and above) conduction of heat by radiation has occurred in test materials, increasing the apparent thermal conductivity. This phenomenon is highly important as a problem in thermal conductivity at high temperatures.

II. CERAMICS THAT EXPLOIT THERMAL PROPERTIES

The thermal properties of ceramics include heat resistance, thermal conductivity, thermal insulation, and low thermal expansion. Since heat resistance is the characteristic property of ceramics, those utilizing their thermal conductivity and low thermal expansion will be discussed here.

A. CERAMICS UTILIZING THERMAL CONDUCTIVITY

1. Thermal Conductivity of Solids

In solids, thermal conductivity can be divided into the conductivities by phonon, photon, and electron. If the specific heats of phonon, photon, and electron are C_1, C_2, C_3, their velocities are v_1, v_2, v_3, and their mean free paths are l_1, l_2, l_3, respectively, then the thermal conductivity k of the solid will be given by Eq. (8.22).

$$k = \frac{1}{3}(C_1 v_1 l_1 + C_2 v_2 l_2 + C_3 v_3 l_3) \tag{8.22}$$

Ceramics are formed of ionic and covalent bonds. Their thermal conductivity is principally affected by phonon transmission through unharmonic reciprocal effects among the lattice vibrations. At high temperatures, however, the contribution of photon conduction by radiation increases, as can be seen in Figure 8.5. Also, in ceramics with low resistivity, the contribution of electron conduction increases. For instance, in ceramics with a resistivity of 100 $\mu\Omega \cdot$cm, the contribution of electron conduction to thermal conductivity is, as calculated according to the Wiedemann–Franz law (Eq. (8.23)), approximately 0.02 cal/s·°C·cm. Thus, the contribution of electron conduction to thermal conductivity cannot be ignored. The carbides, nitrides, and borides of titanium, zirconium, hafnium, and tho-

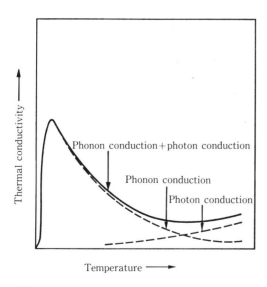

Figure 8.5. Contributions of phonon conduction and photon conduction to thermal conductivity.

rium in the fourth group of the periodic table; vanadium, niobium, and tantalum in the fifth group; and chromium, molybdenum, tungsten, and uranium in the sixth group have small resistivity and in these compounds the contribution of electron conduction to thermal conductivity conduction is large.

$$k_e = \sigma TL \tag{8.23}$$

In Eq. (8.23), k_e is the thermal conductivity due to electron conduction, σ the electrical conductivity, T the temperature, and L the Lorentz number (2.45×10^{-8} W·Ω/K^2).

In ceramics, the resistivity is usually large. For instance, semiconductor ceramics such as SiC and ferrite have resistivities from several to several hundred ohms per centimeter. Insulating ceramics such as Al_2O_3 or forsterite have a resistivity in the range of 10^{14} to 10^{18} Ω·cm. The contribution of electron conduction to thermal conductivity can be ignored in most cases.

2. Thermal Conductivity of Polycrystalline Ceramics

The thermal conductivity of polycrystalline ceramics is affected by the volume fraction, the size, and the shape of the second phase, such as pores. Maxwell and Eucken have reported Eq. (8.24) for the thermal conductivity k_m of the system composed of the continuous phase with thermal conductivity k_c and the dispersion phase with thermal conductivity k_d. In this equation, V_c is the volume fraction of the continuous phase, and V_d the volume fraction of the dispersion phase. In polycrystalline ceramics, the pores are the dispersion phase, and $k_c \gg k_d$. Therefore, the Maxwell–Eucken relation can be closely approximated by Eq. (8.25).

$$k_m = k_c \left\{ \frac{1 + 2V_d(1 - k_c/k_d)/(2k_c/k_d + 1)}{1 - V_d(1 - k_c/k_d)/(k_c/k_d + 1)} \right\} \tag{8.24}$$

$$k_m = k_c[(1 - V_d)/(1 + V_d)] \tag{8.25}$$

Figure 8.6 indicates the relationship between porosity and thermal conductivity for polycrystalline alumina (Touloukian, 1967). The thermal conductivity of polycrystalline alumina increases as the porosity decreases, as is clear from Eq. (8.25).

In polycrystalline ceramics, when impurities exist in a solid solution or as a secondary phase, they reduce the thermal conductivity. That effect is due to phonon scattering by the impurities, thus reducing the mean free path of phonon, which is given by Eq. (8.26).

$$1/l = 1/l_{th} + 1/l_{im} \tag{8.26}$$

In this equation, l is the mean free path of phonon in the system, l_{th} the

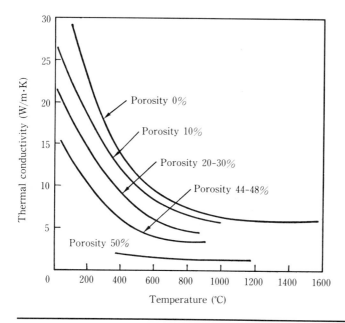

Figure 8.6. Effect of porosity on the thermal conductivity of polycrystalline alumina. From Touloukian (1967).

mean free path determined by thermal scattering, and l_{im} the mean free path determined by scattering by impurities. Figure 8.7 indicates the effect of impurities on thermal conductivity (Kingery et al., 1976). In an NiO-MgO solid solution system, the thermal conductivity of the solid solution is lower than in the pure substance.

The mean free path of phonon is usually 100 Å or less. The thermal conductivity of polycrystalline ceramics is not affected by the grain size in the phonon conduction region. But in the high-temperature region in which photon conduction by radiation contributes to thermal conductivity, the mean free path of photon is on the same order as the grain size. Thus, in that case the grain size does affect the thermal conductivity.

3. Types of Ceramics with High Thermal Conductivity: Their Properties and Applications

Ceramics with high thermal conductivity can be divided into electrical insulating ceramics, in which phonon conduction is dominant, and electrical conductive ceramics, in which electron conduction is dominant.

For ceramics in which phonon conduction is dominant, the following conditions are necessary to produce a high thermal conductivity: The

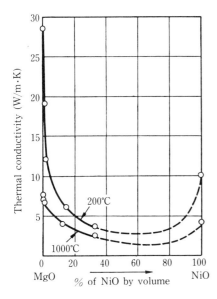

Figure 8.7. Thermal conductivity of the MgO-NiO system. From Kingery, Bowen, and Uhlmann (1976, p. 623).

crystal structure must be simple; the kinds of constituent atoms must be few; the diameters of the constituent atoms must be uniform; the constituent atoms must be light elements; the bond strength between atoms must be high; and pores, impurities, or other defects must be few. For instance, diamond, cubic boron nitride, SiC, BeO, MgO, and Al_2O_3, which satisfy these conditions, are good thermal conductors. In contrast, silicates, including $3Al_2O_3 \cdot 2SiO_2$, $2MgO \cdot 2Al_2O_3 \cdot 5SiO_2$, and stabilized zirconia and quartz glass are poor thermal conductors, due to their complex crystal structures and many kinds of constituent atoms.

Table 8.1 presents a summary of ceramics with high thermal conductivity, their properties, and their applications. Ceramics with high thermal conductivity are in strong demand as materials for electronic components, such as substrates, packages for large-scale integrated circuits to permit higher package density, substrates for high-output semiconductor elements, and heat sinks for laser diodes. In addition to Al_2O_3, insulating SiC and AlN are being developed for these applications (Ura, 1982; Nihon Denshi Kōgyō Shinkōkai, 1983).

B. CERAMICS UTILIZING LOW THERMAL EXPANSION

1. Thermal Expansion of Solids

The atoms or ions composing solids undergo thermal vibration centering at their equilibrium points, determined by the temperature and other factors.

Table 8.1. High Thermal Conductivity Ceramics: Their Properties and Applications

		Thermal conductivity (cal/cm·s·°C)		Melting point (°C)	Thermal expansion coefficient (°C^{-1})	Bending strength (MPa)	Applications
	Crystal structure	100°C	1000°C				
C (Diamond)	Cubic system, diamond type	4.78 (25°C)	—	>3500	3.1×10^{-6}	—	Abrasives and cutting tools
BN	Cubic system, zinc blende type	3.1	—	>3000[a]	4.7×10^{-6}	—	Abrasives and cutting tools
BeO	Hexagonal system, wurtzite type	0.53	0.05	2570	9.0×10^{-6}	240	Headers for power transistors
Al$_2$O$_3$	Hexagonal system, corundum type	0.072	0.015	2050	8.8×10^{-6}	430	Packages and substrates for LSIs; spark plugs
SiC	Cubic (β) and hexagonal (α) systems	0.096	0.030	>2200[a]	4.8×10^{-6}	700	High-temperature structural materials
Si$_3$N$_4$	Hexagonal (α) and hexagonal (β) systems	0.026	0.015	>1900[a]	2.5×10^{-6} (α) 3.5×10^{-6} (β)	1000	High-temperature structural materials
TiN	Cubic system, sodium chloride type	0.016 (500°C)	0.029	2950	9.3×10^{-6}	—	Cutting tools
TiC	Cubic system, sodium chloride type	0.060	0.014	3140	7.4×10^{-6}	1100	Cutting tools
MoSi$_2$	Tetragonal system	0.1 (200°C)	—	2030	7.5×10^{-6}	350	Heating elements

[a] Sublimation.

When one pair of atoms or ions is considered, the potential energy E is the sum of an attraction component and a repulsion component, as can be seen in Eq. (8.27).

$$E = -\frac{A}{r^m} + \frac{B}{r^m} \qquad (8.27)$$

In this equation, r is the distance between a pair of atoms at their equilibrium points and A, B, m, and n are constants. The constants m and n have the relationship $n > m$. Repulsive force shows a sudden increase compared with attractive force, and the resulting potential energy curve is asymmetrical, as shown in Figure 8.8. Therefore, the atoms raise the vibrational energy level to E_1, E_2, E_3 with an increase in temperature. At the same time, the equilibrium position for the atoms is changed to r_1, r_2, and r_3. That is, an increase in temperature causes the distance between atoms to increase and makes the solid substance undergo thermal expansion.

2. The Coefficient of Thermal Expansion in Polycrystalline Ceramics

Thermal expansion coefficients of polycrystalline ceramics can be lowered by exploiting the anisotropy of the coefficients of thermal expansion

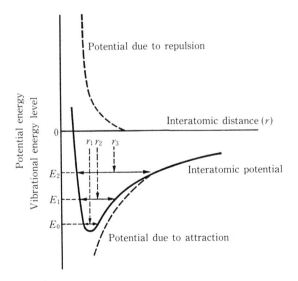

Figure 8.8. *Potential energy curve of a pair of atoms in a solid.*

of the constituent crystals or by forming microcracks at the grain boundaries or in the grains. Cordierite honeycomb ceramics are an example of the former and $Al_2O_3 \cdot TiO_2$ ceramics of the latter.

Since cordierite honeycomb ceramics are used as catalyst carriers to purify automobile exhaust gases, a low thermal expansion is necessary to increase their thermal shock resistance. Cordierite crystal has the thermal expansion characteristic shown in Figure 8.9; the coefficient of thermal expansion in the c axis is negative (Udagawa and Ikawa, 1979). Figure 8.10 indicates the thermal expansion coefficient of each part in the honeycomb ceramics when the c axis of cordierite crystals is perfectly oriented along the thin walls of honeycomb ceramics (Lachman and Lewis, 1975). The coefficient of thermal expansion perpendicular to the thin walls increases to the value of the a axis of the cordierite crystal, but the coefficient of thermal expansion in a parallel direction will be as low as the arithmetic mean value for the a and c axes of cordierite crystal. The high thermal expansion coefficient of honeycomb ceramics in the direction perpendicu-

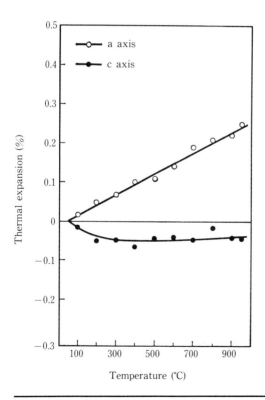

Figure 8.9. Characteristic thermal expansion of cordierite crystals. From Udagawa and Ikawa (1979, p. 967).

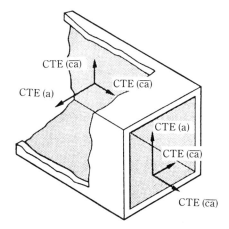

Figure 8.10. Orientation within thin walls of cordierite and the coefficient of thermal expansion for each part of the honeycomb ceramic. From Lachman and Lewis (1975, p. 977).

lar to their thin walls contributes little to the thermal expansion of the whole, due to the specific geometry of the honeycomb structure. That is, the more the c axes of cordierite crystals are oriented along the thin walls, the lower the coefficient of thermal expansion of honeycomb ceramics.

In polycrystalline ceramics, thermal stress may occur, caused by the anisotropic thermal expansion of constituent crystals, by phase transformations, or by the differences in coefficients of thermal expansion among different crystals. This thermal stress may lead to the formation of many microcracks at the grain boundaries and within the grains. In $Al_2O_3 \cdot TiO_2$, $MgO \cdot 2TiO_2$, $HfO_2 \cdot TiO_2$, and Nb_2O_5, there is large anisotropy in the coefficients of thermal expansion for the constituent crystals, leading to the formation of many microcracks and the lowering of the thermal expansion coefficient. Deterioration under use is a problem with $Al_2O_3 \cdot TiO_2$, and research on its prevention is being pursued.

3. Types, Characteristics, and Applications of Low Thermal Expansion Ceramics

Low thermal expansion ceramics must satisfy the following conditions:

1. The electrostatic bond strength is defined as the cation valence divided by the coordination number. The thermal expansion coefficient for the crystal α and the electrostatic bond strength $\mathbf{q}$ have the relationship given in Eq. (8.28). Low thermal expansion ceramics are usually composed of crystals with large electrostatic bond strength ($\mathbf{q}$).

$$\alpha = C/\mathbf{q}^2 (C \doteq 1 \times 10^{-6}) \tag{8.28}$$

Table 8.2. Low Thermal Expansion Ceramics: Their Properties and Applications

Chemical formula	Crystal structure	Coefficient of thermal expansion ($\times 10^{-7}/°C$)	Bending strength (MPa)	Young's modulus (GPa)	Thermal Shock resistance R (°C)	Applications
$2MgO \cdot 2Al_2O_3 \cdot 5SiO_2$	Hexagonal system, beryl type	12	13	16	680	Catalyst carriers for automobile exhaust gases
$Li_2O \cdot Al_2O_3 \cdot 4SiO_2$	Tetragonal system, keatite type	20	70	118	300	Heat exchangers
		10	35	20	1750	Heat exchangers
$Al_2O_3 \cdot TiO_2$	Rhombic system, pseudobrookite type	15	17	10	1130	—
Si_3N_4	Hexagonal system, phenacite type	33	650	230	860	High-temperature structural materials
SiC	Hexagonal system, wurtzite type	43	500	300	380	High-temperature structural materials

2. Low thermal expansion ceramics are usually composed of crystals with strongly covalent interatomic bonds.

3. Low thermal expansion ceramics are usually formed of crystals with loosely packed anions such as oxygen. Crystals that heat-shrink in a specific direction, such as ring silicates or network-structure silicates with a helical axis, may also be used.

Table 8.2 presents low thermal expansion ceramics, their properties, and their applications. Silicon ceramics, such as Si_3N_4, are formed of covalent interatomic bonds, have comparatively small coefficients of thermal expansion, are outstanding in mechanical properties at high temperatures, and offer great promise as engine parts, for instance, as parts for turbochargers or gas turbine engines. Cordierite is also used in highly porous honeycombs for catalyst carriers for automobiles and in dense honeycombs for rotary heat exchangers.

REFERENCES

Kingery, W. D.; Bowen, H. K.; and Uhlmann, D. R. (1976). *Introduction to Ceramics*, 2d ed. New York: John Wiley & Sons.

Lachman, I. M. and Lewis, R. M. (1975). U. S. Patent No. 3,885.

Nihon Denshi Kōgyō Shinkō Kyōkai (ed.). (1983). Shin denshi zairyō ni kansuru chōsa kenkyū hōkoku IX [Research report IX on new materials for electronics]. *Seramikku zairyō chōsa hōkoku, 2.*

Touloukian, Y. S. (1967). *Thermophysical Properties of High Temperature Solid.* Vol. 4, Part 1. New York: Macmillan.

Udagawa, Shigekazu and Ikawa, Hiroyuki. (1979). Teibōchō seramikkusu: netsu bōchō to kesshō kōzō [Low expansion ceramics: Thermal expansion and crystal structure]. *Seramikkusu* [Ceramics, Japan], 14, 967–976.

Ura, Mitsuru. (1982). *Seramikku deetaa bukku '82* [Ceramic data book '82]. Tokyo: Kōgyō Seihin Gijutsu Kyōkai.

9

Chemical Properties

Shigeru Hayakawa and Satoshi Sekido
*Matsushita Research Institute Tokyo, Inc.
Tama-ku, Kawasaki 214, Japan*

I. INTRODUCTION

Chemical reactions proceed by the exchange of elements and electrons between different forms of matter. The chemical properties of ceramics, then, are those that make skillful use of chemical reactions to cause some useful function to take place.

The chemical properties of ceramics may be used in a wide variety of applications. Ceramic chemical sensors, for example, correspond to our senses of smell and taste. They make use of the fact that the electrical qualities of a substance change as the result of exchanges. Another application of the chemical properties of ceramics is chemical batteries and electric double-layer capacitors, which output electric power as the result of a chemical change. Conversely, in chemical pumps, mass transfer is performed as the result of a chemical change effected by the application of an external energy supply, electric current, or light to a ceramic material. Other applications make use of chromism, in which ceramic substances undergo color changes, and electrolytic winning, in which hydrogen and other substances are produced. In addition, there are examples in which ceramics function as catalysts, making necessary chemical reactions run smoothly without being affected themselves.

Space limitations make it impossible to discuss all these chemical applications of ceramics thoroughly. This chapter focuses instead on the categories with the deepest connections to two areas of particular interest, microprocessors and energy, and treats them in detail.

II. CHEMICAL SENSORS

Of all the substances in the external world, it is necessary for a chemical sensor to respond to only the particular substance for which it is targeted. The development of sensors, then, has been a struggle for selectivity. These sensors include gas sensors, for flammable gas or humidity, and ion concentration sensors.

There are, roughly speaking, three approaches to increasing sensor selectivity (Seiyama, 1982, p. 5; Seiyama et al., 1982, p. 14). One approach, used in ceramic humidity sensors, uses adsorption. Another uses chemical reactions; it is used in semiconductor gas sensors, surface electrometer gas sensors, and catalytic combustion-type gas sensors. Finally, there is the use of permselective membranes, as in oxygen gas concentration and ion concentration sensors.

First we consider adsorption. When moist gases are adsorbed by a sensor substrate, dissociative adsorption (chemical adsorption) occurs on the surface of the substrate, and relatively weaker bonding (physical adsorption) occurs on the outside of the dissociative layer. It is thought that, if we consider the adsorption of moisture in terms of bonding strength, then OH^- adsorption is stronger than H^+ adsorption, which is stronger than physical adsorption; these three types of adsorption occur in that order from the inner side of the sensor surface to the outside (Arai, 1982, p. 40; Seiyama, et al. 1982, p. 91). One type of humidity sensor that has reached the stage of practical use exploits changes in resistance caused by physical adsorption and desorption of water. In these sensors the substrates are ceramics such as $MgCr_2O_4$ or $MgFe_2O_4$. These ceramics are stable at high temperatures, and changes in their resistance due to adsorption and desorption can be readily detected. Adding a small amount of TiO_2 is said to be effective in increasing sensitivity in the low-humidity region. In these sensors, the sensor substrate is also wrapped in kanthal wire heaters for heat cleaning, to prevent changes over time. In cooking, in which taste and aroma control are fundamental, there had been a trend to avoid using chemical sensors because of reliability problems. A humidity sensor, however, has the advantage of more readily detecting boiling than can a temperature sensor. This point has led to its being widely used for control of cooking times in microwave ovens (Nitta and Terada, 1976, p. 885; 1980, pp. 433 and 450).

Now we look at sensors that use chemical reactions. A semiconductor gas sensor uses a sintered body of an oxide, such as SnO_2, ZnO, or TiO_2, as a substrate. By setting the operating temperature, the oxidation reaction of the flammable gas is selected and the gas concentration is detected by changes in resistance. In general, the logarithm of the electrical resistance is proportional to that of the gas concentration (Yamazoe, 1982, p. 31; Seiyama, 1982b, p. 30). Generally, the change of resistance depends on that of gas concentration. However, as the grain size of the oxides used is made finer, there is a grain-size region in which the carrier mobility is dependent upon the gas concentration. With grain size fine enough to obtain that effect, sensitivity can be raised (Ogawa, 1981, p. 80). This type of gas sensor is used widely to detect gas leaks of flammable gases.

One type of surface electrometer gas sensor has Pd attached as a gate electrode for a field-effect transistor (Lundström, Svensson, and Bakowski, 1977a, 301; Lundström, Shivaraman, and Svensson, 1977b, p. 245; 1975, p. 3876). Another surface electrometer gas sensor is a junction diode comprised of Pd and an oxide such as TiO_2 (Yamanoto, Tonomura, Matsuoka, and Tsubomura, 1980, p. 400). Then, when H_2 and CO appear and react with the oxygen adsorbed by the Pd, the surface potential of the Pd, that is, the threshold voltage of the field-effect transistor and the junction potential of the junction diode change. The sensor uses these effects. Such sensors are still at the research stage, but they have been attracting interest for their potential for miniaturizing sensors.

In catalytic combustion-type gas sensors, platinum wire resistors are buried in a highly porous alumina carrier. One side has a Pt or Pd catalyst attached and one does not. The catalysts are connected to the two arms of a bridge and a balance resistors is reached when there are no flammable gases. When an flammable gas does appear, the temperature of the side with the catalyst rises, the balance is lost, and an electric output current in proportion to the gas concentration is produced (Arakawa, Adachi, Shiokawa, 1982, p. 24; Seiyama, 1982b, p. 73).

All flammable gases will react with the precious-metal catalyst. Since the oxidation reaction of flammable gases with large polarity, such as alcohol, is fast, this type of sensor has the defect of making false reports with cooking alcohol. To avoid this problem, one can compensate by putting a Cu catalyst, which reacts only with alcohol, in the carrier which does not have the precious-metal catalyst (Kubota, Sukigara, Akiyama, Sakurai, Fukuda, Shikae, 1981, p. 75). This application has aroused considerable interest, for it indicates one direction for using reactions in sensors.

Finally, there are sensors using either organic polymers or solid electrolytes as permselective membranes. The distinctive feature of these sensors is the combination of electrochemical measurement methods, potentiometry and amperometry, in detection. When this type of sensor uses solid electrolytes, it is regarded as belonging to the field of ceramics.

Solid electrolytes are impervious to all but electrically conductive ions, a characteristic used to raise their selectivity. Potentiometry is a means of learning the concentration of a gas or of ions by forming a concentration cell and measuring the electromotive force produced. For example, consider a zirconia electrolyte oxygen sensor. The zirconia electrolyte is ZrO_2 to which is added 8 mol% Y_2O_3 or 9 mol% MgO; it is then fired. The crystals are in fluorite form with positive ions as the lattice points in the face-centered cube and O^{2-} ions in the positions on the four faces formed by those positive ions. Due to replacement of Zr^{4+} by Y^{3+} or Mg^{2+}, those positions are empty here and there, giving rise to O^{2-} ion conductivity. Platinum black is applied as the electrodes to both sides of a sintered board of this electrolyte. One side is exposed to an atmosphere in which the oxygen concentration has a definite value such as that of air ($P''_{O_2} = 0.21$ atm) and the other is exposed to the atmosphere for which the concentration is to be measured. Then, assuming the oxygen partial pressure in the atmosphere to be measured is P'_{O_2}, a concentration cell is formed, and electromotive force is produced. This electromotive force is expressed by Eq. (9.1).

$$E = RT/4F \cdot \ln [P'_{O_2}/P''_{O_2} \text{ (constant)}] \tag{9.1}$$

Here, R is the gas constant and F is Faraday's constant, so that if the temperature T is held constant, the oxygen concentration can be found by measuring the electromotive force. That is the principle underlying this method.

Although a variety of substances may be present, a concentration cell is formed only by O^{2-} ions, which are electrically conductive ions, the O_2 gas in an equilibrium relationship with them, and metal ions. This type of sensor has been designed in a variety of forms; it is used to control the air/fuel ratio in automobile engines and to control the O^{2-} ion concentration in molten metals such as iron or copper.

The form of sensor sketched in Figure 9.1 is used to control automobile engine air/fuel ratios. The inner electrode is exposed to air, the outer electrode to the exhaust gases. Oxidized gases such a NO_x and O_2 are mixed with reduced gases such as CO and hydrocarbons in the exhaust. These gases react with each other due to the catalytic action of the platinum in the outer electrode. The electromotive force produced is dependent upon the oxygen concentration left after the reaction has been completed. Because the oxygen concentration in the exhaust gas changes suddenly when the fuel ratio crosses the stoichiometric composition point, the electromotive force also changes greatly at that point, as shown in

Figure 9.1. If the air/fuel ratio is controlled to the stoichiometric composition point, not only will the fuel efficiency of the engine rise; the system also has the additional advantage that, by passing the exhaust after combustion through a catalyst layer, the harmful gas content can be minimized (Mizusaki, Yamauchi, Fueki, 1982, p. 46; Seiyama *et al.*, 1982, p. 123).

A sensor of the same construction is used for achieving a fixed quantity of oxygen ions in molten metals. The differences are that instead of having air as the reference electrode, a mixture of the metal iron or chromium

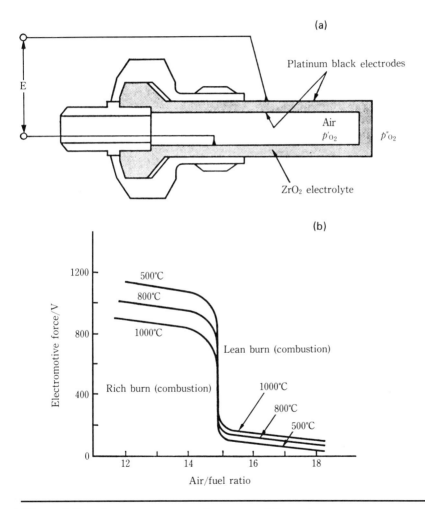

Figure 9.1. An oxygen sensor for automobile engine control (a λ sensor). (a) shows the structure of a λ sensor and (b) shows the relationship between the air/fuel ratio and the operating voltage of the λ sensor.

with their oxides is used. In addition, the molten metal itself is taken as the outer electrode and a molybdenum or iron lead is immersed in it (Gotō, 1982, p. 54).

Potentiometric sensors using a solid electrolyte are also used to measure ion concentrations in water. In this case, part of the container is formed of a solid electrolyte membrane, as shown in Figure 9.2, and the container is filled with a solution of known ion concentration. An electrode consisting of a silver wire coated with silver chloride (called the Ag/AgCl electrode) is immersed in that solution. This assembly as a whole is regarded as one

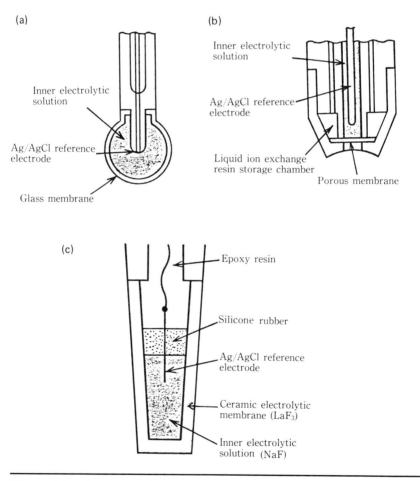

Figure 9.2. Structure of ion concentration sensors [ion selective (membrane) electrodes]. (a) shows the glass membrane electrode, (b) the liquid ion exchange membrane electrode, and (c) the LaF_3 membrane electrode.

electrode and is called the ion selective electrode. The ion selective electrode and the other Ag/AgCl electrode are immersed in the solution which is to be measured, and the electromotive force generated across the two electrodes is measured.

A variety of ion selective electrodes have been developed, as Table 9.1 shows. Since, if the electrolytic membrane used in these electrodes is dissolved the measured values will be affected, materials should be chosen less for high ionic conductivity than for being hard to dissolve. These membranes may be of vitriform, sintered form, and combined form with silicone rubber, and single crystals of insoluble ionic materials. Since the ion species with which the conductive ion has an equilibrium relationship is not necessarily of one type, those other than the ion to be measured are interfering ions. If the concentrations of the ion to be measured and the interfering ions are C_A and C_B and their electrovalences are Z_A and Z_B, then the electromotive force E is given by Eq. (9.2).

$$E = RT/Z_A F \ln (C_A + \kappa C_B{}^{Z_A/Z_B}) \tag{9.2}$$

where κ is the selectivity coefficient, which expresses the proportion of interfering ions. The ions measured, the electrolytes used, the upper and lower limits of measurement, and the selectivity coefficient are presented in Table 9.1. (Durst, 1969; Moody and Thomas, 1977; and Niki, 1982, p. 13).

Amperometry is also used in chemical sensors. Figure 9.3 presents an example of a sensor under study for maintaining the automobile engine

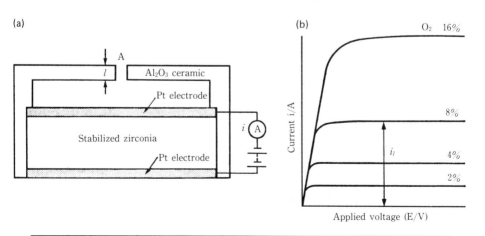

Figure 9.3. *Measurement of oxygen concentration through amperometry. (a) shows the structure of the measurement cell, and (b) shows the relationship between voltage and current.*

Table 9.1. Properties of Major Ion Selective Electrodes

Ion species	Membrane composition	Measurement range	Operating temperature	pH range and response time	Interfering ions (selectivity coefficient)[a]
Na$^+$	Glass NAS$_{11-16}$	$\approx 10^{-6}$	0–60°		Ag$^+$ > H$^+$ > Na$^+$ > Li$^+$ > K$^+$ > NH$_4^+$ Na/K 2800, H/Na 35.7
K$^+$	NAS$_{27-5}$KAS$_{20-5}$	$\approx 10^{-6}$	0–60°		H$^+$ > Ag$^+$ > K$^+$, NH$^+_4$ > Na$^+$ > Li$^+$ K/Na 20
Ag$^+$	MAS$_{28.8-19}_{11-18}$	$\approx 10^{-6}$	0–60°		Ag/H 10^5 Ag/Na 10^3
	Precipitate-impregnated crystal membrane				
S^{2-}	Ag$_2$S	10^0–10^{-17}	0–50°	~pH 12 1 min	CN$^-$
F$^-$	LaF$_3$, NdF$_3$, PrF$_3$	10^0–10^{-6}	—	–pH 3	OH$^-$, Al^{3+}, Fe^{3+}
Cl$^-$	AgCl/Ag$_2$S[b]	10^0–5 × 10^{-3}	0–50°	pH 3–11 2 sec	CN/Cl 5 × 10^6,[b] I/Cl 2 × 10^4 NO$_3$/Cl 5–9, Br/Cl 3 × 10^2
Br$^-$	AgBr/Ag$_2$S	10^0–10^{-5}	0–50°	pH 1.5–11.5 2 sec	CN/Br 1.5 × 10^4,[b] I/Br 5 × 10^3
I$^-$	AgI/Ag$_2$S	10^{-1}–10^{-7}	0–50°	pH 2–11 30 sec	CN,[b] S^{2-}, MnO$_4^-$
Ag$^+$	Ag$_2$S	10^0–10^{-17}			Hg^{2+}
CN$^-$	[c]	10^{-2}–10^{-6}			S^{2-}, I$^-$, MnO$_4^-$
SCN$^-$	AgSCN/Ag$_2$S	10^0–10^{-5}			Strong reducing agents
SO$_4^{2-}$	BaSO$_4$	10^{-1}–10^{-6}			Cl$^-$ (>10^{-3} M)

Ion	Membrane	Concentration range (M)	Temp/Response	pH range	Selectivity (interferences)
Zn^{2+}	Mixed-crystal membrane ZnS + Ag_2S	10^{-1}–10^{-6}	0–50°		$Na/Cu\, 5 \times 10^{-4}, Fe/Cu\, 140, Hg^{2+}, Ag^+$
Cu^{2+}	CuS + Ag_2S	10^{-1}–10^{-7}			Ag^+, Cu^{2+}, Hg^{2+}
Cd^{2+}	CdS + Ag_2S	10^{-1}–10^{-7}			Ag^+, Cu^{2+}, Hg^{2+}
Pb^{2+}	PbS + Ag_2S				
Divalent cations	A salt with greater solubility than Ag_2S		30 sec		
Cl^-	Ionic radical of liquid ion exchange membrane $-NR_4^+$	10^{-1}–10^{-5}		pH 2–11	$ClO_4/Cl\, 32, I/Cl\, 17, NC_3/Cl\, 42$
NO_3^-				pH 2–12	$I/NO_3\, 20, ClO_4/NO_3\, 10, ClO_3/NO_3\, 2$
BF_4^-					$I/BF_4\, 20$
	$[Fe("\ ")]^{2+}$				
ClO_4^-		10^0–10^{-5}		pH 4–10	$OH/ClO_4, 10$
Cu^{2+}	$R\text{-}S\text{-}CH_2\text{-}CO_2^-$	10^{-1}–10^{-5}		pH 4–7	$Fe/Cu\, 1400, H/Cu\, 10$
Pb^{2+}		10^{-2}–10^{-5}			$Cu/Pb\, 26$
Ca^{2+}	$(RO)_2PO_2^-$	10^0–10^{-5}		pH 3.5–7.5	$H/Ca\, 10^7, Zn/Ca\, 32$
Ca Mg		10^0–10^{-8}		pH 5.5–11	$Zn\, 35, Fe\, 35, Cu\, 21, Ni\, 135$

[a] The value which expresses the effect of the interfering ions on the electrode that has as its target the analysis of the selected ion is derived from Eq. (9.2).

[b] If an Ag_2S membrane is made on the surface, this is neglected.

[c] If Ag_2S membrane is used, CN^- ion concentration is taken at twice what is given by the Nernst equation.

air/fuel ratio in the lean burn region. The electrode consists of platinum applied to both sides of a sintered wafer of zirconia solid electrolyte. The negative pole is covered with Al_2O_3 ceramic with small pores. When D. C. voltage is applied to this sensor, the oxygen is gathered in by the negative pole, where it becomes oxygen ions. These transfer to the positive pole and are released as oxygen gas. The transfer speed is proportional to the current flow. When the applied voltage is raised, the oxygen gas in the negative pole chamber will gradually become less concentrated and eventually reach zero. At that saturation point, raising the applied voltage further will not increase the current.

This saturation current is the limiting current. It corresponds to the state in which the diffusion velocity of the oxygen from the outside, having passed through the small pores, has reached its limit. This limiting current can be expressed in the form of Eq. (9.3).

$$i_l = 4FDAP_{O_2}/l \tag{9.3}$$

In this equation, D is the diffusion constant, A the cross section of the pores, and l their length. In this way, with amperometry one measures the limiting current to learn the concentration, but both have a proportional relationship. Therefore, the detections sensitivity is higher than in potentiometry (Takeuchi and Igarashi, 1981, p. 1).

III. CHEMICAL BATTERIES AND ELECTRIC DOUBLE-LAYER CAPACITORS

Since it is possible to prevent the occurrence of other than the necessary reaction when using the ion selectivity permeability of a solid electrolyte, it is possible to devise batteries and electric double-layer capacitors with little self-discharge and a high charging efficiency. In addition, a solid electrolyte with high ion conductivity is itself an ionic crystal. When an electrical field below its decomposition voltage is applied, the conductive ions are set in motion and are deposited at one electrode. Then ions with the opposite charge remain in high concentration around the opposite electrode, opposing the charge on the electrode at a molecular order distance. Thus, a capacitor with a large capacitance is formed.

In both batteries and capacitors, it is preferable to have high-current charging and discharging. At room temperature, however, the only high-conductivity electrolytes are, at present, Ag^+, $Cu+$, and $H+$ ion conductors, and their decomposition voltage is low. Therefore, at room temperature such devices cannot be obtained with high terminal voltages by using high-conductivity electrolytes.

A. BATTERIES FOR HIGH-TEMPERATURE USE

The battery types of particular interest are the Na–S battery and the coal-gas fuel cell. A Na–S battery has the structure shown in Figure 9.4. In this secondary battery, a fused sodium negative electrode in a closed cylinder of β'-or β''-alumina solid electrolyte is immersed in fused sulfur containing fibers of graphite and stainless steel. Its characteristics are its voltage (about 2 V) and high energy density, at about 150 Wh/kg, approximately five times that of a lead-acid storage battery. It is being studied for use in electric vehicles or in load-leveling applications (Kummer and Weber, 1967; Nippon denshi Kōgyō shinkō kyōkai, 1980). Beta alumina is expressed by the chemical formula $NaAl_{11}O_{17}$; it has a layered structure with conductive layers of Na^+ ions of 4.76 Å between 11.23 Å spinel blocks. It has been discovered that, when part of this Al^{3+} is replaced by Mg^{2+}, the spinel blocks expand to 16.8 Å and, to preserve electrical neutrality conditions, a large number of conductive Na^+ ions are admitted. This alumina is expressed by the chemical formula $Na_{1.67}Mg_{0.67}Al_{10.33}O_{17}$ (Yao and Kummer, 1967, p. 2453; Weber and Kummer, 1967, p. 37).

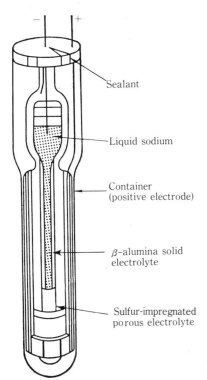

Figure 9.4. Structure of a Na–S battery.

Figure 9.5 presents the structure of a coal-gas fuel cell being studied. On a tube of porous zirconia are placed a fuel electrode of a mixture of Ni and zirconia, a connector of $CoCr_2O_4$ to which has been added a small amount of Mn, and an air electrode of In_2O_3 with a small admixture of Sn or of $Sr_xLa_{1-x}MnO_3$. These are attached by sequential flame spraying and vacuum evaporation, so that batteries connected in series are formed on a single tube. These batteries using coal gas can generate electricity at the high conversion efficiency of more than 60%, at an estimated cost of 0.4 cents per kilowatt-hour. This performance is superior to the 40% efficiency and 0.9 cents per kilowatt-hour of the newest steam power generation system. Since the equipment costs are less than $30 per kilowatt-hour and the fuel cells produce little air pollution, their potential has been the subject of increasing interest (Nippon denshi kōgyō shinkō kyōkai, 1980, p. 42; 1981, p. 122).

B. BATTERIES FOR ROOM-TEMPERATURE USE

Since an electrolyte with a high decomposition voltage and high conductivity at room temperature cannot be obtained, it is difficult to produce a high-voltage solid state battery with a high drain capability. The spread of the use of C–MOS electronic circuits, however, has created demand for a battery with a high energy density even with a slight load current. Solid electrolyte batteries with a lithium negative electrode have, for example, already proved their usefulness in heart pacemakers or to back up volatile memory.

Two types of materials are used as the the positive electrode in such batteries: PbI_2 and the iodine complex, including poly 2 (vinylpyridine) (Moser, 1972a, p. 163; 1972b, p. 562) and n-butyl pyridinium polyiodide (Sekido, Nakai, Sotomura, 1978, p. 52; 1979a, p. 976; 1979b, p. 240; Takahashi et al., 1980, p. 51; Whittingham, 1981, p. 263). Batteries using

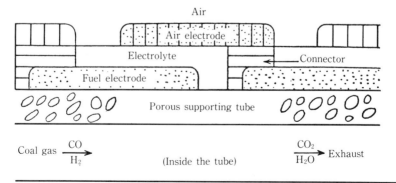

Figure 9.5. Structure of a coal-gas fuel cell.

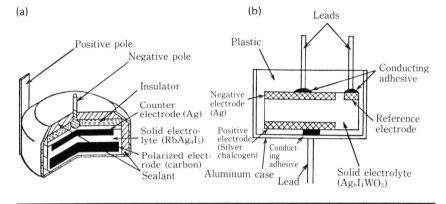

Figure 9.6. Elements of electric double-layer capacitors. (a) is a cross-sectional view of an ESD element, and (b) shows an analog memory element.

an iodine-complex positive electrode have an open-circuit voltage of 2.8 V. That type of battery is always made with the active material of the positive electrode pressure-welded to the lithium plate of the negative electrode. Batteries using PbI_2 have an open-circuit voltage of 1.95 V and are designed with an electrolytic layer of $2LiI \cdot Li_3N \cdot 0.77LiOH$ (Kawakami, Wada, Mochizuki, Uetani, Kudō, 1981, p. 153; Asai, Nagai, and Kawai, 1982, p. 75) or $6LiI \cdot 4Al_2O_3$ (Liang, 1973, p. 1289). Both types have a low self-discharge rate, less than 5% over 10 years.

C. ELECTRIC DOUBLE-LAYER CAPACITORS

As Figure 9.6 (a) indicates, these capacitors use an electrolyte of $RbAg_4I_5$ (Oxley, 1970, p. 20) or $RbCu_4I_{1.5}Cl_{3.5}$ (Sekido, Ninomiya, 1981, p. 153; Nippon denshi kōgyō shinkō kyōkai, 1981, p. 56). The polarized electrode that forms the double-layer capacitor uses activated charcoal or metallic chalcogen, and Ag or Cu + Cu_2S is used for the opposite electrode, depending on the type of electrolyte. The two types can withstand voltages of 0.76 and 0.6 V, respectively. These capacitors can easily achieve a farad-order capacitance, with little current leakage. An additional feature is their long life span under repeated chargings and dischargings.

IV. CHEMICAL PUMPS

As is explained in Figure 9.3, when platinum electrodes are attached to both sides of a zirconia electrolyte and direct current is applied, oxygen transfers from the negative to the positive electrode compartment. Their

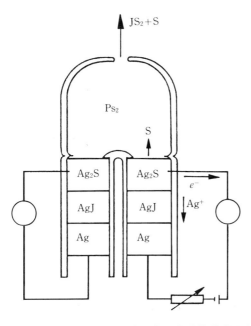

Figure 9.7. Structure of an electrochemical Knudsen cell.

velocity is proportional to the current drained. Only the oxygen transfers, even if there are many gases at the negative electrode compartment. This pump is used to pump oxygen into the atmosphere, to prevent the multiplication of anaerobic bacteria, or to pump out oxygen from the atmosphere, to prevent decomposition by aerobic bacteria (Nippon denshi kōgyō shinkō kyōkai, 1980, p. 11; 1981, p. 88). Similarly, as shown in Figure 9.7, in chemical vapor deposition, emission velocity of the vapor of a substance (the example in the figure uses S vapor) is controlled by electric current (Gotō, 1974, p. 11; Rickert, 1972, p. 524).

Since this use of current to effect mass transfer chemically is similar to the operation of a pump, these devices have been named chemical pumps. Chemical pumps have the advantages that the substances which will transfer when a solid electrolyte is used as the electrolyte are limited to only those which are associated with electrically conductive ions and that the efficiency of transformation is high.

V. ELECTROCHROMISM

The production of color by adding electrical energy is called electrochromism. Examples include color changes centering on the alkali halide colors due to cathode rays (Schulman and Compton, 1962, p. 52), the coloring of

Mn and Fe in $BaTiO_3$ due to charge transfer (Blanc and Staebler, 1971, p. 3548; Staebler, 1975, p. 177), and the coloring due to the electrolytic deposition of silver from silver halide. Considerable research has been carried on in the field for some time, but the example closest to practical application is the gain and loss of color due to the electrolytic deposition and elimination of H^+, Li^+, Na^+, and Ag^+ ions on WO_3 (Beni, 1981, p. 157). This reaction is as given in Eq. (9.4)

$$WO_3 + x Me^+ + xe^- \underset{\text{(losing color)}}{\overset{\text{(developing color)}}{\rightleftharpoons}} Me_x WO_3 \text{ (blue)} \quad (9.4)$$

The reaction occurs in proportion to the current drained and is reversible. Figure 9.8 shows its basic structure. For a display terminal, WO_3 is applied by vapor deposition to a thickness of 1,000 to 3,000 Å to a transparent electrode with indium oxide on glass. There are two types of electrolytes, liquid and solid. In the liquid type, sulfuric acid was used initially but, for longer life, lithium salts dissolved in propylene carbonate or γ-butyrolactone came to dominate. For the solid type, those such as Li_3N, β-alumina, or $RbAg_4I_5$, which have the requisite conductive ions to produce color in WO_3, are being considered. If a solid electrolyte is used, it is possible to limit the types of conductive ions and to increase the efficiency of the electrode reaction. Since high-temperature displays are possible with solid electrolytes, they look quite promising. To prevent differences in the darkness of the display, it is desirable for the counter electrode to have a stable electrical potential and for the reaction to be reversible. A variety of materials are under consideration.

Displays using WO_3 have a number of points of superiority over existing liquid crystals and light-emitting diodes. Their operating voltage is low, and they consume little electrical power. They possess a display memory function. They do not luminesce, but their readability is not affected by the angle of vision. The display segments form electrochemical couples, which, since they possess electromotive force, make a readout simple. Applications which take advantage of these characteristics are expected to develop.

VI. HIGH-TEMPERATURE STEAM ELECTROLYSIS

Thanks to the energy problem, research into the storage of electrical and solar energy by hydrogen has been booming. Hydrogen can be stored in almost the same volume as its liquid form by alloys of the La-Ni and Ti-Mn families. When it is burned in that state, no harmful exhaust gases are produced, and it is possible to reproduce the electrical energy at a high degree of efficiency using the fuel cell illustrated in Figure 9.5.

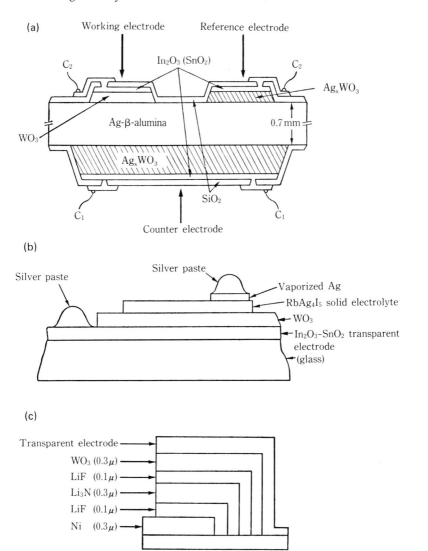

Figure 9.8. Electrochromic displays (ECDs) using solid electrolytes. (a) shows an ECD using a β-alumina solid electrolyte, (b) shows an electrochromic display using an $RbAg_4I_5$ solid electrolyte, and (c) shows an electrochromic display using an Li_3N solid electrolyte.

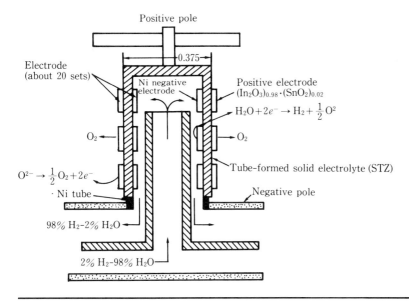

Figure 9.9. *The GE new type of steam electrolysis.*

The electrolysis of water has long been the conventional method used to produce hydrogen. Interest is now focusing, however, on the steam electrolysis technique shown in Figure 9.9, because it has the advantage of not having bubbles develop on the electrodes and because, due to the high temperatures involved, the voltage loss is extremely low. Production costs for 1,000 standard cubic feet per second are estimated to be about 25.2 cents, in contrast to 33.6 cents by the conventional high-pressure electrolysis method (Tamura, 1974, p. 212).

Research is under way on photochemical electrolysis of water by exposing a semiconductor electrode to light, but that approach has disadvantages in terms of efficiency and corrosion of the electrodes. In contrast, a technique utilizing perovskite-type oxides such as $SrTiO_3$ has begun to draw notice (Bard, 1982, p. 172; Kerchove, Vandermolen, Gomes, and Cardon, 1979, p. 230; Viswanathan, Narayanan, Viswanath, and Varadarajan, 1982, p. 199).

VII. CATALYSTS

To find some way to eliminate the harmful exhaust gases produced by automobile engines and combustors is critical for the fight against air pollution.

Precious-metal catalysts and oxide catalysts had been used previously. Oxide catalysts are inexpensive, but they can be used only to do away with the reduced gas CO and hydrocarbons and it is difficult to eliminate NO_x. Precious-metal catalysts are called *three-way catalysis* for being able to eliminate NO_x as well if there is an excess of reduced gases, but they have the disadvantage of high price.

Recently, exploration has turned to the possibility of using perovskite-type oxides such as $SrFeO_3$ (Shin, et al., 1982, p. 199), $Sr_xLa_{1-x}MnO_3$ (Johnson et al., 1976, p. 520), and $Sr_{1+x/2}La_{1-x/2}Co_{1-x}Fe_xO_3$ (Tachibana, Yamamura, and Sekido, 1982, p. 53; Nakamura, Misono, Uchijima, and Yoneda, 1980, p. 1679) for eliminating harmful gases. These compounds, without an alteration in their crystalline structure, give oxygen from their lattices to reduce gases and take into their lattices the oxygen from oxidized gases. The effect is readily obtained and is expected to be useful in catalyst applications.

In chemical applications, ceramics have been used principally as solid electrolytes. Their selective permeability is employed not only in chemical sensors but also in batteries and electrolytic application devices that previously had used conventional liquid electrolytes. These ceramics are contributing to the conversion to solid state systems and the increase of current efficiency in such devices.

In addition, electrolytes with high ion conductivity have a high ion density and low mobility, so that it is possible to use them in electric double-layer capacitors with a high charge density and good memory. The substances which have been developed to have high ion conductivity at room temperature have a low decomposition voltage, however, and even if their physical properties are utilized, the magnetic, optical, and thermal interrelations are virtually an uncharted field. The potential for developments in these directions in the near future is great.

REFERENCES

Arai, Hiromichi. (1982). Shitsudo sensā [Humidity sensors]. *Denki Kagatzu* [Journal of the Electronic Society of Japan] 50, 38–45.

Arakawa, Tsuyoshi; Adachi, Ginya; and Shiokawa, Jirō. (1982). Sesshyoku nenshōshiki gasu sensā [Gas sensors using resistance change by catalytic combustion sensors]. *Denki Kagtzu* [Journal of the Electronic Society of Japan] 50, 24–28.

Asai, Takeshi, Nagai, Tatu; and Kawai, Nanao. (1982). Richiumu ion dōdentai $Li_3N \cdot 2LiI \cdot 0.77LiOH$ chū no Li^+ ion no kakusan [The dispersion Li^+ ions in the lithium ion conductor $Li_3N \cdot 2LiI \cdot 0.77LiOH$]. In *Kotai ionikusu tōronkai* No. 9 yokō [Preprint of No 9 meeting on solid ionics] pp. 75–76.

Bard, A. J. (1982). Design of semiconductor photoelectrochemical systems for solar energy conversion. *Journal of Physical Chemistry*, 86, 172–177.

Beni, G. (1981). Recent advances in inorganic electrochromics. In M. S. Whittingham (ed.), *Solid State Ionics 3/4*. Amsterdam: North-Holland, pp. 157–163.
Blanc, J. and Staebler; D. L. (1971). Electrocoloration in strontium titanate: vacancy drift and oxidation-reduction of transition metal. *Physics Review B*, 4, 3548–3577.
Durst, R. A. (1969). *Ion selective electrodes*. Washington: NBS Standard Special. Publication 314.
Gotō Kazuhiro. (1974). Kotai denkaishitsu ni kansuru shinpojiumu [Symposium on solid electrolytes]. Tokyo: Nihon Kinzoku gakkai.
Gotō, Kazuhiro. (1982). Tekkō seirenyō sanso sensā [An oxygen sensor used in steel refining]. *Denka*, 50, 54–63.
Johnson, D. W., Jr.; Gallagher, P. K.; Schrey, F.; and Rhodes, W. W. (1976). Preparation of high surface area substituted $LaMnO_4$ catalysts. *American Ceramic Society Bulletin*, Vol. 55 no. 7, 520–523.
Kawakami, Akira; Wada, Shūichi; Mochizuki, Shōji; Uetani, Yoshio; and Kudō, Tetsuichi. (1981). Richiumu-yōkanamari denchi no hōden tokusei [The discharge characteristics of lithium-lead iodide batteries]. In *Denchi tōronkai No. 22 yokō* [Preprint of No. 9 meeting on batteries], 153–155.
Kerchove, F. V., Vandermolen, J.; Gomes, W. P.; and Cardon, F. (1979). Investigation on kinetics of electroreduction processes at dark TiO_2 and $SrTiO_3$ semiconductor electrode. *Berichte der Bunsen-Gesellschaft für Physikalische Chemie*, 83, 230.
Kubota, Issei; Sukigara, Kunio; Akiyama, Shigeru; Sakurai, Kazuo; Fukuda, Tokuyuki; and Shikae, Shunsuke. (1981). Toshi gasu sensā [Sensors for urban gas utilities]. In *Sensaa no kiso to ōyō shinpojiumu* [Proceedings of the 1st sensor symposium of the Institute of Electrical Engineering of Japan], Tokyo: Denki gakkai, pp. 75–76.
Kummer, J. T. and Weber, Neill. (1967). Sodium-sulfur secondary battery. *SAE paper*, No. 670179.
Liang, C. C. (1973). Conduction characteristics of the lithium-iodide-aluminum oxide solid electrolyte. *Journal of the Electrochemical Society*, 120, 1289–1295.
Lundström, K. I., Shivaraman, M. S., and Svensson, C. M. (1975). A hydrogen-sensitive Pd-gate MOS transistor. *Journal of Applied Physics*, 46, 3876–3880.
Lundström, K. I., Shivaraman, M. S., and Svensson, C. M. (1976). Hydrogen in smoke detected by Pd-gate field effect transistor. *Review of Scientific Instruments*, 47, 738–740.
Lundström, K. I., Shivaraman, M. S., and Svensson, C. M. (1977). Hydrogen sensitive MOS structure. *Vacuum*, 27, 245–247.
Lundström, K. I., Svensson, C. M., and Bakowski, M. (1977). The use of Pd-MOS transistors to monitor Hz, H_2S and NH_3. In *30th ACEMB*, November 5–9, *301*.
Mizusaki, Junichirō; Yamauchi, Shigeru; and Fueki, Kazuo. (1982). Kotai denkaishitsu sensā no kangaekata; Kotai denkaishitsu gasu sensā [View of solid electrolyte sensors; solid electrolyte sensors]. *Denka*, 50, 7–12; 46–53.
Moody, G. J. and Thomas, J. D. R. (Munemori Makota, *et al.*, Tr.) (1977). *Ion sentaku denkyoku* [Selective ion sensitive electrodes]. Tokyo: Kyoritsu Shuppan.
Moser, J. R. (1972a). U.S. Patent No. 3,660,163. Lithium iodide battery.
Moser, J. R. (1972b). U.S. Patent No. 2,674,562. Lithium iodide battery.

Nakamura, Teiji; Misonō, Makoto; Uchijima, Toshio; and Yoneda, Sachio. (1980) Perobusukaitokei fukugō sankabutsu no sanka taiō kassei [Activity in response to oxidation of perovskite-type complex oxides] *Nikka*, 1679–1684.

Niki, Eiji. (1982). Denki kagakuteki ionshu shikibetsu [An electrochemical classification of ion types]. *Denka*, 50, 13–16.

Nippon denshi kōgyō shinkō kyōkai. (1980). Kotai denkaishitsu zairyō chōsa hōkoku 1 [Research report on solid electrolytic materials 1]. In *Shindenshi zairyō in kansuru chōsa kenkyū hōkokusho VI* [Report on the investigation of new electronic materials]. Tokyo: Nippon denshi kōgyō shinkō kyōkai.

Nippon denshi kōgyō shinkō kyōkai. (1981). Kotai denkaishitsu zairyō chōsa hōkoku 2 [Research report on solid electrolytic materials 2]. In *Shindenshi zairyō ni kansuru chōsa kenkyū hōkokusho VII* [Report on the investigation of new electronic materials]. Tokyo: Nippon denshi kōgyō shinkō kyōkai.

Nitta, Tsuneharu and Terada, Jirō. (1976). Seramikku kanshitsu teikōtai soshi [Ceramic humidity-sensitive resistance devices]. *National Technical Report*, 22, 885–894.

Nitta, Tsuneharu and Terada, Jirō. (1980). Takinō sensā 'seramikku shitsudo-gasu sensā' hyumiseramu III [A multifunctional ceramic humidity-gas sensor, Humiceram III]. *National Technical Report*, 26, 450–456.

Ogawa, Hisahito. (1981). Sankasuzu chōbi ryūshi maku no bussei narabi ni gasu sensā e no ōyō [Tin oxide complex ultrafine particle membranes: their physical properties and applications in gas sensors]. Doctor's thesis, n.p.

Oxley, J. E. (1970). Solid state energy storage device. *Proceedings of Power Sources Symposium*, 24, 20–23.

Rickert, H. (1972). Kinetic properties of the galvanic cell. In J. Hladik (ed.), *Physics of Electrolytes*. New York: Academic Press, pp. 524–534.

Schulman, J. H. and Compton, W. D. (1962). *Color Centers in Solid*. Oxford: Pergamon.

Seiyama, Tetsurō. (1982) Gasu sensā kitai seibun no denki kagakuteki shikibetsu [An electrochemical classification of the gaseous constituents of gas sensors]. *Denki Kagaku*, 50, 2–6.

Seiyama, Tetsurō *et al.* (ed.) (1982). Kagaku sensaa [Chemical sensors]. Tokyo: Kōdansha.

Sekido, Satoshi; Nakai, Muneaki; and Sotomura, Tadashi, (1978). Richiumu-yōso sakutai denchi no busshitsu [Lithium-iodine complex battery substance]. In *Denchi tōronkai yokō* [Preprint of 19th meeting on batteries], No. 19, pp. 52–53.

Sekido, Satoshi; Nakai, Muneaki; and Sotomura, Tadashi. (1979). U.S. Patent No. 4, 148,976. Litnium-iodine complex solid eletrolytic battery.

Sekido, Satoshi; Nakai, Muneaki; and Sotomura, Tadashi. (1979b). Richium-yōso sakutai kotai denkaishitsu denchi [Lithium-iodine complex solid electrolytic battery]. In *Denki kagaku taikai yokō,* No. 46, [Preprint of 46th meeting on electrochemistry], pp. 240–241.

Sekido, Satoshi and Ninomiya, Yoshito. (1981). Solid state electric double layer elements using Cu^+ ion conductive solid electrolyte. In M. S. Whittingham (ed.), *Solid State Ionics 3/4*. Amsterdam: North-Holland, pp. 153–156.

Shin, Shigemitsu; Arakawa, Hironori; Yonemura, Michiko; Ogawa, Kiyoshi; Kiyosumi, Yoshimichi; Suzuki, Kunio; Shimomura, Kinya; and Ikawa,

9. Chemical Properties 187

Hiroyuki. (1982). Kōonyō seramikkusu shokubai (fukisoku sanso kekkan perobusukaito ni yoru NO no netsu bunkai taiō) [High-temperature ceramic catalyst (Thermal decomposition function of NO by irregular oxygen defect perovskite)]. *Shokubai*, 24, 34–36.

Staebler, D. L. (1975). Oxide optical memories: photochromism and index change. *Journal of Solid State Chemistry*, 12, 177–185.

Tachibana, Hirokazu; Yamamura, Yasuharu; and Sekido, Satoshi. (1982). Perobusukaitokei kongō dōdentai no shokubai sayō [Catalytic function of perovskite type mixed conductor]. In *No. 9 kotai ionikusu tōronkai yōko* [Preprint of 9th meeting on solid ionics]. Tokyo: Kotai ionikusu garubaruni denchi kenkyūkai, pp. 53–54.

Takeuchi, Takashi and Igarashi, Isemi. (1981) Jidōsha yō sanso sensā [An oxygen sensor for the automobile use]. In *Kagaku sensaa kenkyu happyōkai* [Preprint of 1st meeting on chemical sensors], 1, 1–6.

Tamura, Hideo. (1974). Suiso energī shisutemu to denkikagaku [The hydrogen energy system and electrochemistry]. In *No. 16 Denkikagaku seminā* [16th electrochemical seminar, Kansai division of the Electrochemical Society of Japan]. Osaka: Denkikagu kyokai Kansai shibu, pp. 199–235.

Tonomura, Shōichirō; Takechi, Satoshi; Matsuoka, Tsuguhumi; Yamamoto, Naoto; and Tsubomura, Hiroshi. (1980). Kinzoku-handotai setsugōgata gasu sensā to sono sadō kikō [Metal semiconductor composite gas sensors and their operational structures]. *Nikka*, 1585–1590.

sensā to sono sadō kikō [Metal semiconductor composite gas sensors and their operational structures]. *Nikka*, 1585–1590.

Viswanathan, B., Narayanan, S. R., Viswanatn, R. P., and Varadarjan, T. K. (1982). *Indian Journal of Technology*, 20, 199.

Weber, Neill and Kummer, J. T. (1967). Sodium-sulfur secondary battery. In *Proceedings of the Annual Power Sources Conference*, 37–39.

Whittingham, M. S. (ed.) (1981). Reaction mechanisms in the Li/LiI/1-n-butyl pyridinium polyiodide solid electrolyte cell. In *Solid State Ionics 3/4*. Amsterdam: North-Holland, pp. 263–266.

Yamamoto, Naoto; Tonomura, Shōichirō; Matsuoka, Tsuguhumi; and Tsubomura, Hiroshi. (1980). A study on a paradium-titanium oxide Schottky diode as a detector for gaseous components. *Surface Science*, 92, 400–406.

Yamazoe, Noboru. (1982). Handōtai gasu sensā [Semiconductor gas sensors]. *Denki Kagku*, 50, 29–37.

Yao, Y. Y. and Kummer, J. T. (1967). Ion exchange properties of and rate of ionic diffusion in β-Al_2O_3. *Journal of Inorganic Nuclear Chemistry*, 29, 2453–2475.

10

Optical Properties

Toru Kishii
Toshiba Glass Co., Ltd.
Haibara-gun, Shizuoka 421-03, Japan

Hitoshi Hirano
Toshiba Corporation
Saiwai-ku, Kawasaki 210, Japan

I. NONCRYSTALLINE SUBSTANCES

A. GLASS LASERS

Glass lasers are made of glass containing Nd^{3+} ions that perform fluorescent-center functions. The glass is formed into a rod with the two ends parallel. A semitransparent, reflective film is applied by vapor deposition to one end and a reflective film to the other. When the center is stimulated by ultraviolet or shortwave visible light, the Nd^{3+} ions mutually release induced fluorescent light. Finally, as it travels back and forth between the reflective surfaces, the light in phase comes to predominate (see Figure 10.1) and part of it passes through the semitransparent end as laser light.

The light produced by a glass laser is in short pulses of some tens of nanoseconds wide. The instantaneous power output is great, so that the laser can be used in machine tools, such as those for welding or cutting. Large lasers with large outputs are readily obtained. These are used in

Note: Toru Kishii wrote section I; section II is by Hitoshi Hirano.

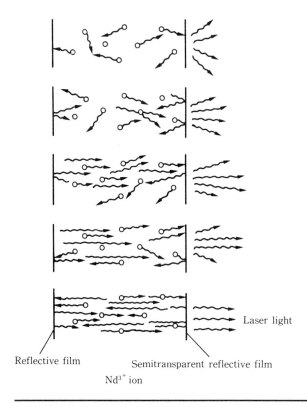

Figure 10.1. Process within glass containing Nd^{3+} ions excited by an external light source until the laser beam is emitted (top to bottom).

research on nuclear fusion, in which the fusion reaction is caused by irradiation and implosion of heavy hydrogen enveloped in spheres.

In a glass laser, increasing the laser light output may damage or break the glass. The output limit depends upon the composition, production method, uniformity, and optical characteristics of the glass. For instance, with a 30 nsec pulse width, the internal damage threshold is 400 J·cm^{-2} and the external damage threshold is 80 J·cm^{-2} (13 and 2.6 GW·cm^{-2}, respectively). A glass-fiber laser has proved successful in experimental testing.

B. GLASS FOR LASER AMPLIFICATION

In terms of the qualities of materials used, the same glasses are used in laser amplification as in glass lasers. The glass is exposed to light, the Nd^{3+} ions are pumped up in it until they reach an excited state, and the beam of the glass laser is passed through this glass. The stimulated emission of radiation from the excited ions causes them to emit light which matches the laser light in phase and direction. This emitted light then moves on with the incident laser light (Figure 10.2). Experimental devices for nuclear fusion reactions use lasers with multistage amplifiers.

C. FARADAY ROTATION GLASS

In optical memory and reading technology, it is necessary, for light intensity modulation or one-way passage of light, to combine linearly polarized light with rotation of the plane of polarization of the light. Faraday rotation is one of the methods for rotating the plane of polarization.

The glass is placed in a magnetic field and linearly polarized light is shone on it; then, as the light passes through the glass and the magnetic field, the plane of polarization will rotate (Figure 10.3). If another polarizing sheet is placed in the incident light, the light transmitted will be a function of the angle of rotation. The angle of rotation θ is proportional to the magnetic field H and the distance l the light passes through the glass, as given in Eq. (10.1):

$$\theta = VHl \tag{10.1}$$

The proportional constant V is determined by the glass used; it is called Verdet's constant. Verdet's constant for glass available commercially has reached $0.09 \, \text{min} \cdot \text{Oe}^{-1} \cdot \text{cm}^{-1}$ at a wavelength of 633 nm and $0.24 \, \text{min} \cdot \text{Oe}^{-1} \cdot \text{cm}^{-1}$ at a wavelength of 1,060 nm.

Glasses with a large Verdet's constant are known to have a large content of PbO, Bi_2O_3, TeO_2, and other oxides of metals with high atomic numbers or to have a large content of oxides of rare earths. The relationship between the magnetic field and the direction of rotation of the plane of polarization is opposite for these two types of glass.

Applications of Faraday rotation glass include telemetry of a high-voltage line current by the incidence of linear polarized laser light through the medium of the magnetic field surrounding the line.

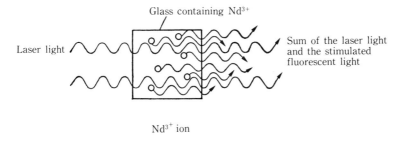

Figure 10.2. Process by which glass containing Nd^{3+}, excited by external light, amplifies a laser beam.

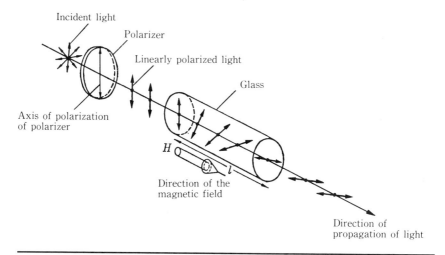

Figure 10.3. Faraday rotation of glass in a magnetic field H.

D. GLASS FOR ACOUSTOOPTIC ELEMENTS

Techniques for making laser light deflect and for scanning are needed in memory, read-out, display, and printer applications.

Use of acoustooptic element glass is a simple means of obtaining high resolution and response speed in such applications. A piezoelectric element is used to excite elastic waves within the glass. When that happens, compressive elastic waves are produced at equal intervals of density and index of refraction distribution (Figure 10.4). These waves act as a diffraction grating, causing the production of diffracted light from the incident laser light. The angle of diffraction is determined in response to the frequency of the elastic waves. Light moving straight ahead and unneeded diffracted light are excluded, and only one type of diffracted light is used in scanning.

The commercial use of this principle in facsimile printers has become widespread. For highly efficient devices, glass with a high index of refraction and a large change in the index of refraction accompanying stress is required. Glass with TeO_2 as the principal constituent has become commercially available.

E. PHOTOCHROMIC GLASS

The special quality of photochromic glass is that it changes color due to ultraviolet or shortwave visible light and returns to its original color when that light is cut off. Used in sunglasses, the glass darkens in a bright environment, when sunshine strikes it, and clears when in the shade or

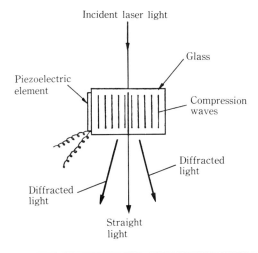

Figure 10.4. Diffraction of light by compressive elastic waves within glass.

indoors. The color changes require minutes, but improvements year after year have speeded up the response time, with an accompanying increase in sales. Experimental use in automobile sunroofs has also been reported.

Several types of glass are known to have a photochromic property: glasses containing Eu^{2+} ions, thallium halide crystals, or silver halide. Only the silver halide glass is used commercially.

Silver halide is dispersed within the glass as crystalline colloidal fine particles several hundred angstroms in size. Light causes the following reaction to proceed to the right side, and heat (including room temperatures) and red light cause the reverse reaction to proceed to the left side.

$$Ag^+X^- \underset{\text{Heat or longwave light}}{\overset{\text{Shortwave light}}{\rightleftarrows}} Ag^0 + X^0$$

X : Halide

Accompanying the change from Ag^+ to Ag^0, the colloidal particles absorb visible light.

An example of the darkening when light strikes and losing color when the light is cut off is presented in Figure 10.5. The speed of the color change and the sensitivity of the glass make it usable only in eyeglasses at the present time.

F. OPTICAL FIBERS

Glass fibers in a medium with a low index of refraction, such as air, or double-structure fibers (with the inner high refraction index glass surrounded by a low refraction index glass) act to transmit light without

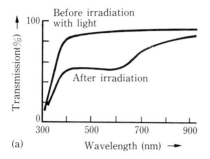

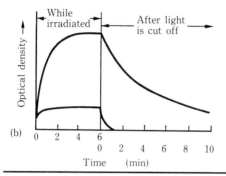

Figure 10.5. Examples of photochromic glass response speeds for (a) acquiring color and (b) acquiring and losing color.

divergence. Thus, such fibers are used to provide lighting for gastrocameras and devices for internal inspection of pipes.

Recent improvements in glass composition and higher purity had indicated that the possible transmission distance could be extended to kilometer scale distances. That prediction has now come true. When glass fibers are used in optical communications, a signal can be transmitted up to 50 km without a repeater. Other applications of the same technology include measurement, processing, and control activities, for which large-diameter fibers delivering a large output of light have been marketed.

Optical fibers are categorized in terms of their materials as the silica or quartz glass group (some of which contain small amounts of additives) and the multicomponent glass group. In terms of their structure, they can also be divided into glasses of a step-index type, those of a graded-index type, and those of a single-mode type (Figure 10.6).

Propagation of light within the fibers is explained as presented in terms of ray optics in Figure 10.6. Since the light is moving ahead, enclosed in a region of width close to its wavelength, however, a wave optical effect

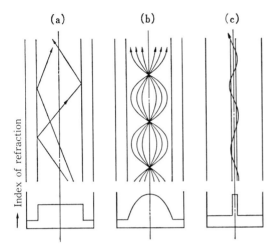

Figure 10.6. Refractive index distributions (lower diagram) and transmission paths, represented in terms of ray optics (upper diagram). (a) shows the step-index type fiber, (b) the graded-index type fiber, and (c) the single-mode type fiber.

appears, so that, of the possible paths given by ray optics, only those which correspond to a finite number will be permitted and will occur. Those permitted paths are called the modes. As the number of conducting paths decreases, the mode number decreases. Finally there will be only one mode (the single-mode type).

In the quartz glass group, the highly pure glass body that is to be converted to the wave path is synthesized by the oxidation or pyrolysis of a silicon compound gas. The multicomponent group of glasses are formed by the fusion in a platinum crucible of materials refined to a high level of purity.

The quartz glass group is suitable for long-distance transmissions since it has an extremely low level of light losses. Those losses, presented in Figure 1.7 as functions of the wavelength, are at present nearing their theoretical lower limit. Multicomponent glasses are readily formed into high-input fibers. With glass with an index of refraction gradient or with a single mode it is simple to obtain fibers with a wide transmission bandwidth.

Communication via optical fiber, including optical data transmission, is now a reality. These communications systems have wide bandwidths and send on multiple circuits and multiple channels through the use of wavelength- and time-multiplexing technology. There is a degree of freedom in system design. The optical fibers are resistant to high temperatures and humidity and do not suffer electromagnetic faults or interference.

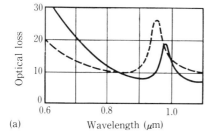

(a)

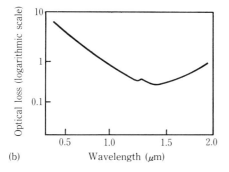

(b)

Figure 10.7. Relationship between losses and wavelengths of light in optical fibers. (Losses of up to about 20 dB can be transmitted without repeaters.) (a) shows a multicomponent glass fiber of a graded-index type and (b) shows a quartz glass fiber.

While fiber optic systems possess all these strengths, only a part of their potential has been harnessed thus far.

1. Optical Fiber Products

Due to progress in techniques to reduce transmission losses, the possible transmission length of fiber bundles for visual transmission use has grown. Since the transmittance of the visible light and infrared has improved, optical pyrometers and radiation pyrometers of the type which measure temperatures by transmitting the radiation from a far-distant subject have become a reality.

With a refractive index gradient applied to the interior of a glass rod, the rod acts like a concave or convex lens with a wide-range focal length, due to the depth of the cross section (Figure 10.8). Thus, it is possible to use such glass rods in optical instruments, in condensing lenses for optical semiconductors, and in collimating lenses for optical fiber systems. A set of

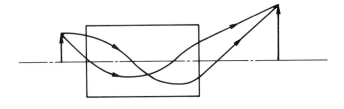

Figure 10.8. Diagram of a graded-index rod performing the same image formation function as a convex lens.

these rod lenses, each about 1 mm in diameter, lined up lengthwise, has two-dimensional image scanning and image formation capabilities. Used in cameras, such lenses are contributing to the miniaturization and simplification of their construction.

2. Optically Conductive Chalcogenide Glass

Sulfur, selenium, and tellurium are the chalcogenide elements. In combination with arsenic, antimony, or other metals they readily form amorphous substances. These amorphous substances are formed by sudden cooling of the solution or by the thermal vapor deposition of the raw materials and compounds.

These amorphous substances conduct both electricity and light. Moreover, heat crystallizes them, changing their electrical conductivity considerably. The latter capability can be regarded as acting as a switch, and extensive research has been carried out to bring it to practical use. At present, however, there are no examples of its successful commercialization. In contrast, the optical conductivity of these substances has made them an essential element in major industrial products such as copiers and video imaging tubes.

In the manufacture of copiers, selenium is applied by vapor deposition to a metal drum. An electric charge is placed on the surface of the selenium by a high-voltage electrical discharge (Figure 10.9). The drum revolves in synchronization with the scanning of the material to be copied, and light is shone on the selenium surface of the drum, causing image formation there. Where the light strikes it, the selenium becomes electrically conductive, and the electric charge disappears. If brought into contact with a black pigment powder, the powder will be held by static electricity to the parts of the drum on which the charge remains, that is, the parts which correspond to the parts of the original which were not blank. Thus, an image just as on the original will be reproduced there. That image in powder, then, is transferred to paper and fixed there, creating the copy as output.

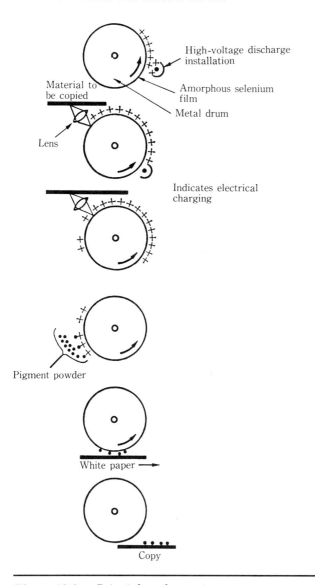

Figure 10.9. Principles of a copier using a noncrystalline selenium film. (The flow of time is from top to bottom.)

A vidicon, or video imaging tube, has a light-sensitive photoconductive film. The film may be made of PbO, PbS, amorphous chalcogenides, or crystalline chalcogen compounds. The photoconductive film is made over a film of In_2O_3, which is transparent and electrically conductive (Figure 10.10). The image is projected on the photoconductive film, and an elec-

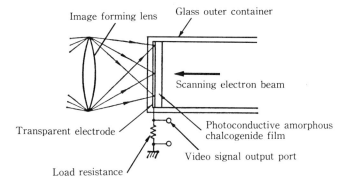

Figure 10.10. Operation of a vidicon video imaging tube using a photoconductive amorphous chalcogenide film.

tron beam scans the film, whereupon the differences in electrical resistance in various parts of the film are converted externally to currents of electricity of various strengths, making the television signal. Video imaging tubes made with an amorphous chalcogenide are known under the trademark Saticon.

II. CRYSTALS

The optical properties of crystals have been applied intensely since the discovery of lasers in the 1960s. A devoted effort was made to develop functional devices using crystals, to make practical use of laser light, which is ultrashort wavelength electromagnetic waves. The many attempts include applications of several physical phenomena, for example, the Pockels effect and the Kerr effect, which had been known previously.

The optical characteristics of crystals capable of application in optical devices can be divided into two categories. One includes absorption and luminescence phenomena. The other is linked to the refractive index. The first category of characteristics has become well known through applications in the laser beam oscillations using yttrium aluminum garnet (YAG) and ruby crystals. In what are known as optically functional crystals, the second category of characteristics comes into play; applications of optically functional crystals, such as KDP and $LiNbO_3$, use their changing refractive index. In crystals with these characteristics, interesting functions such as secondary harmonic generation and optical rectification, related to a higher order refractive index (nonlinear effect), have been realized. Recently, diffusing impurities into such crystals has also led to the realization of devices that, as optical wave guides, performing laser beam modulation on the wave guide.

An extensive range of optical applications of crystals was known even before the discovery of lasers. Classic examples are the polarizers and wave plates ($\frac{1}{4}$ wave plate, $\frac{1}{2}$ wave plate) for microscopes. The crystals used in such applications, however, were for the most part natural crystals such as quartz, calcite, and mica, in contrast to the manmade crystals used after the discovery of lasers. In fact, the development of techniques for growing artificial crystals has gone hand-in-hand with advances in lasers. This linked development has stimulated the growth of a new field, the application of crystals with optical properties.

We first discuss basic optical properties and then describe in detail the characteristics of frequently used crystals and examples of their applications, to give a better understanding of the optical functions of crystals.

A. BASIC OPTICAL PROPERTIES

The laser effect was first achieved using a ruby crystal in 1960. The principle of the laser is shown in Figure 10.11 (Shimoda et al., 1972). When electrons excited by some external stimulus to high energy levels drop to lower energy levels, they emit ultrashort wavelength electromagnetic waves (light). The emission may be spontaneous or induced. If the laser crystal is placed in a resonator and almost all the emitted light is contained in a feedback loop, the induced emission is amplified. Part of the emitted light becomes output light, in the form of loss from the resonator; that light

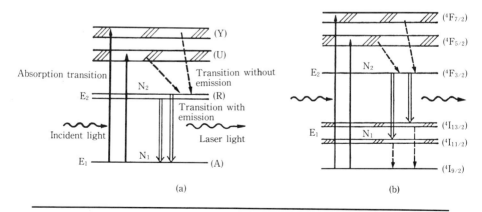

Figure 10.11. Principle of the solid laser. (a) shows a three-level laser. Y, U, R, and A correspond to the ruby absorption lines. (b) shows a four-level laser. The $^4F_{7/2}$, $^4I_{9/2}$, etc. on the right correspond to the level of YAG: Nd^{3+}

is the laser beam. For amplification to occur, the number of electrons N_2 at energy level E_2 must be greater than the number N_1 of electrons at energy level E_1 (Figure 10.11). That is, a so-called negative temperature must be formed. Negative temperature formation is readily obtained for solids with the three-level laser, typified by the ruby laser, and the four-level laser, such as crystals of YAG with a dopant of Nd, known as a continuous-wave oscillation solid-state laser.

When crystals are insulators, the refractive index for interaction with light is only the real part (n) of the complex refractive index (Ariyama et al., 1958). When the internal energy of the crystal is represented in the form of Helmholtz free energy (A), taking the second derivative of the polarization (P_i) gives this refractive index (the real part of the complex refractive index) as the reciprocal of the dielectric constant ($1/\varepsilon_{ij} = 1/n_{ij}^2$) (Mitsui, 1969).[1] The Helmholtz free energy A can be expressed by a higher order polynomial expression for a thermodynamic variable such as χ_i (strain) or the electric field (E_i). Thus, ($1/n_{ij}^2$) can similarly be expressed by a higher order polynomial expression. For E, the first-order term γE and the second-order term γEE express the Pockels effect (γ) and the Kerr effect g; and the coefficients $\gamma \equiv \gamma_{mi}$ and $\mathbf{g} \equiv g_{mm}$ are the third- and forth-rank tensors. Also, as interaction between the light and the crystal, if P is generated within the crystal with incident light E^ω and, due to P^ω, the output light E^ω is emitted, then $P^\omega = \eta E^\omega$, where η is equivalent to n. When E^ω is large, η becomes $\eta = \eta_0 + \mathbf{d}E^\omega + \ldots$, including nonlinear terms.

Table 10.1 summarizes several interactions of light and crystals from this viewpoint. In it, **d** is a tensor of the same rank as γ, known as the second harmonic generation coefficient. As is shown in the Table, γ is the coefficient when n increases in proportion to E, and **g** is the coefficient when n increases in proportion to E^2; and, as the variable n changes, the modulation of light is also indicated by these coefficients. **d** expresses, in the form $\mathbf{P}^{\omega l} = \eta_0 E^\omega + \mathbf{d}E^\omega E^\omega$, that in $\mathbf{P}^{\omega l}$ is included harmonic content which varies in the form $E^{2\omega}$.

The tensor components of a crystal are closely related to the symmetry of the crystals. The existence of each non-zero tensor component is determined uniquely by the 32 crystal classes, into which crystals are classified, corresponding to the point group elements that express their symmetry (Nye, 1960). For instance, all crystals have a non-zero component for η_0 and **g**, but for γ and **d**, non-zero components exist only for 20 point-group elements without a center of symmetry.

The combination of the components is determined by the crystal orientation, and the vector components E, P, and so on form a new vector

---

[1] The subscripts i, j, k, l are the integers 1, 2, 3; m and n are the integers 1, 2, ..., 6; m (and similarly n) are combined with (ij) and (kl) as $1 \rightarrow (1, 1), 2 \rightarrow (2, 2), 3 \rightarrow (3, 3), 4 \rightarrow (2, 3), 5 \rightarrow (1, 3)$, and $6 \rightarrow (1, 2)$.

Table 10.1. *Polarization within Crystals and Optical Characteristics*

Polarization	Optical characteristics of crystals
$P_i^\omega = \sum_{j=1}^{3} \varepsilon_{ij} E_j^\omega$	Dielectric constant (inverse of refractive index) Uniaxial crystals $\varepsilon_{11} = \varepsilon_{22} \neq \varepsilon_{33}$ Biaxial crystals $\varepsilon_{11} \neq \varepsilon_{22} \neq \varepsilon_{33}$
$P_i^\omega = \sum_{j=1}^{3} \sum_{k=1}^{3} \gamma_{ijk} E_k^0 E_j^\omega$	Pockels effect
$P_i^\omega = \sum_{j=1}^{3} \sum_{k=1}^{3} \sum_{l=1}^{3} g_{ijkl} E_l^0 E_k^0 E_j^\omega$	Kerr effect
$P_i^\omega = \sum_{j=1}^{3} \sum_{n=1}^{6} q_{ij,n} S_n E_j^\omega$	Photoelastic effect S_n is strain (also written S_{kl}), a combination of $n \rightarrow (kl)$
$P_i^\omega \sum_{j=1}^{3} \sum_{k=1}^{3} \sum_{l=1}^{3} l_{ijkl} B_l^0 B_k^0 E_j^\omega$	Cotton-Mouton effect B_l^0 and B_k^0 are magnetic flux density
$P_i^{2\omega} = \sum_{j=1}^{3} \sum_{k=1}^{3} d_{i,jk} E_j^\omega E_k^\omega$	Second harmonic generation

component. That result expresses the interaction of the crystal with light. Therefore, in applying the optical properties of a crystal, it is necessary, while selecting the appropriate crystal, to achieve the optimal combination of the crystal orientation and the components E and P.

B. CRYSTALS WITH OPTICAL PROPERTIES

Of laser crystals, the outstanding example of crystals with light-emitting functions, the best known are rubies ($AlO_3:Cr^{3+}$) and YAG ($Y_3Al_5O_{12}:Nd^{3+}$) (Maiman, 1960, p. 493; Geusic, 1964, p. 182). Ruby crystals have aluminum in coordination number 6 in close-packed hexagonal oxygen lattices. The additive Cr^{3+} replaces the Al^{3+}. At that point, the d-shell electrons in the iron family ion Cr^{3+} are split off by the crystal field and form an energy level structure in which is observed, as an absorption spectrum, with two R lines in the region of 690 nm, three B lines near 480 nm, and broad U and Y bands at 550 and 407 nm. Their absorption is due to the $d-d$ transition, and the red coloration of the crystal is due to the electron transition from R to the ground level A. Electrons excited from A to U and Y make the transition to R without emitting light; R to A is the light-emitting transition. This energy level determines the structure of a three-level laser. In addition, the wide bandwidth of the U and Y levels is favorable for excitation.

When the rare earth Nd^{3+} is added to YAG, light emission corresponds to the f–f transition of f shell electrons. When the emission center is the rare earth, the crystal field of the matrix substance has almost no effect upon the energy level. Therefore, the emitting wavelength is determined by the type of additive ion. If Nd^{3+} is added to glass or $CaWO_4$ crystals, the same wavelength, 1.06 μm, is found as with YAG. Thus, the lowest excitation level ($^4F_{3/2}$) is the absorption line near 900 nm and the ground level (^4I) is a quadruplet, including the three levels $^4I_{9/2} < {}^4I_{11/2} < {}^4I_{13/2}$, which contribute to the transition. The light emission at transition is most intense (1.06 μm) in the $^4F_{3/2} \rightarrow {}^4I_{11/2}$ transition. The energy difference ($^4I_{11/2} - {}^4I_{9/2}$) within ^4I is an order of magnitude greater than the thermal energy at room temperature. Then the negative temperature between $^4F_{3/2}$ and $^4I_{11/2}$ is readily formed, with electrons hardly staying. Therefore, crystals containing Nd^{3+} are excellent media for four-level lasers. In particular, YAG is a crystal for solid continuous laser applications with outstanding characteristics, including its crystal quality.

Many crystals have been developed which are suitable for interaction with light mediated through the refractive index, for example, the second harmonic generation. Almost all are ferroelectric crystals (Jona and Shirane, 1962). Outstanding examples are KDP, ADP, $LiNbO_3$ (LN), and $LiTaO_3$ (LT).

The compound KDP (KH_2PO_4) is ferroelectric below 122 K, as is ADP ($NH_4H_2PO_4$) below 148 K. At room temperatures, they are known as piezoelectric crystals, and belong to the point group $\bar{4}2\,m$. In this crystal, the Pockels coefficient is a zero component except for $\gamma_{41} = \gamma_{52}$, γ_{63} (Mitsui, et al. 1969). The Kleiman rule, however, gives $\mathbf{d}$; $\mathbf{d}_{41} = \mathbf{d}_{52} = \mathbf{d}_{63} \doteqdot 5 \times 10^{-13}_{m/V}$. There is only one type of numerical value for $\mathbf{d}$. The refractive index corresponds to that of a negative uniaxial crystal, and the ordinary refractive index (n_0) is greater than the extraordinary refractive index (n_e). For visible light, these refractive indices indicate normal dispersion.

Both LN and LT belong to the trigonal system. At less then about 1210 and 665°C, respectively, they are ferroelectric with a $3m$ point group. They are uniaxial crystals with four γ types: $\gamma_{22} = -\gamma_{12} = -\gamma_6$, $\gamma_{13} = \gamma_{23}$, $\gamma_{51} = \gamma_{42}$, and three nonzero components for $\mathbf{d}$ since $\mathbf{d}_{13} = \mathbf{d}_{51}$. If $\mathbf{d}_{41} = 1$ for KDP, then the three values of $\mathbf{d}$ for LN are $\mathbf{d}_{22} \doteqdot 6$, $\mathbf{d}_{13} \doteqdot 12$, and $\mathbf{d}_{33} \doteqdot 83$, one order of magnitude or more greater. Again, LN has slightly larger values for γ: In contrast to $\gamma_{63} = 1 (10.3 \times 10^{-12}$ m/V and $\gamma_{41}/\gamma_{63} \doteqdot 0.8$ for KDP), γ for LN are $\gamma_{22} \doteqdot 0.3$, $\gamma_{51} \doteqdot 2.7$, $\gamma_{13} \doteqdot 0.8$, and $\gamma_{33} \doteqdot 2.9$.

The actual performance of the light modulation, given by a function of the refractive index and γ, is about 8.3 times greater for LN, when the γ_{51} of LN and the γ_{63} of KDP are compared. Nonetheless, in contrast to a $\gamma_{63} = 1$ for KDP, ADP has $\gamma_{63} \doteqdot 0.8$ and $\gamma_{41} = 2.3$, which is slightly larger, although the refractive indices are almost the same. In comparisons of LN and LT, LT is slightly smaller in the refractive index and in γ, but the effective light modulation is about the same.

Apart from those crystals, quartz, $KNbO_3$ (Takei et al., 1969; Koide et al., 1969, p. 1516), $Ba_2NaNb_5O_{15}$, and others have been used in the laboratory. Quartz, however, has a small coefficient and it is difficult to grow high-quality crystals of the others. Thus, KDP and LN predominate. Also, KTN ($KT_a(1 - \chi) Nb_x O_3$) has attracted attention since it is a crystal with a large Kerr effect. It presents difficulties in terms of crystal quality, including changes in composition, that have kept it from practical use.

To realize the outstanding characteristics of materials in optical devices, it is necessary to grow high-quality crystals of appropriate size. Among the various techniques of crystal growth (Pamplin, 1975), the CZ method (also known as the Czochralski method or the crystal pulling method) has progressed furthest in terms of improvements in equipment and technique; it is now an established method for growing crystals. In this method, seed crystals are touched, while being rotated, to a liquid raw material melted at high temperatures. Then the crystal is pulled from the supercooled layer on the surface of the liquid. Latent heat escape from the jig holding the seed crystal and the surface of the crystal being pulled up drives crystal growth; and controlling the pulling speed and the temperature of the liquid gives a controlled crystal diameter.

The raw materials are melted under heat in a crucible. The melting points of YAG, LT, and LN are 2050, 1650, and 1253°C, respectively. Given the high temperatures involved, YAG and LT are usually melted in an iridium crucible, while a platinum crucible is used with LN. Platinum is lacking in reactivity at high temperatures and in air, but iridium will oxidize. Thus, iridium is used in an inert gas atmosphere. Initial materials used include $LiCO_3$ and Nb_2O_5 for LN, $LiCO_3$ and Ta_2O_5 for LT, and Al_2O_3 and Y_2O_3 for YAG. Resistance heating and induction heating are both used in melting LN. Induction heating is used with LT and YAG. In all cases, to obtain crystals of high quality, it is necessary to use good-quality raw materials, to control changes in the growing conditions, and at the same time to keep the liquid-solid interface in equilibrium, controlling the convection of the melt, to maintain a small temperature gradient just above the melt surface and a small pulling rate.

Rubies, also crystals with a high melting point, can also be grown by the CZ method, but their growth by the Verneuil method is better known. In the Verneuil method, also called the flame fusion method, powdered materials are passed through the high-temperature flame of an oxyhydrogen burner, melted, and deposited at the tip of a heat-resistant rod, promoting crystallization. Accompanying the crystallization, the rod is pulled down and growth at the tip of the crystal is observed at a fixed position, controlling the flame and the quantity of the raw material. The disadvantage of this method is that thermal distortions in the crystals grown in this way make them crack easily. In addition, if the grains of the raw materials are large or if the growth rate is fast, cracks and air bubbles easily form in the crystals.

In contrast, KDP and ADP are grown by the solution method (*Kesshō kōgaku handobukku*, 1971, p. 810). A saturated aqueous solution of the water-soluble crystal materials, for which the solubility possesses a positive temperature coefficient, is gradually cooled. Accompanying cooling, the supersaturated solute is deposited, and crystals grow on the seed crystals suspended in the mother liquid. The growth rate is dependent upon the solubility temperature coefficient and the rate of cooling, but it is possible to grow large-scale crystals in a short time. To eliminate factors which reduce crystal quality, including, for example, impurities, striae, pores, spontaneous seed crystals, and splinters, it is necessary to prepare high-quality materials and seed crystals and equipment with few fluctuations in growing conditions and to use the optimal pH of the mother liquid and optimal growth speed.

C. APPLICATIONS

Now consider a structure of a light modulator such as shown in Figure 10.12, in which the right-angled parallelepiped LN crystals with their edges parallel in the x, y, and z directions are sandwiched between a mutually crossed polarizer and an analyzer. The y–z plane of the crystals is parallel to the surface of the polarizer and analyzer, and the polarizer's direction of polarization is at 45° angle to the y and z axes. When a light beam is transmitted into the crystal along the x-axis direction (the crystal length l), the electric field component of the beam is $E_y^\omega = E_z^\omega$ in front of the crystal. In this arrangement, since refractive indices corresponding to E_y^ω and E_z^ω are n_0 and n_e ($n_0 \neq n_e$), respectively, the light beam enters the analyzer with a phase difference

$$\Gamma = 2\pi (n_0 - n_e)l/\lambda$$

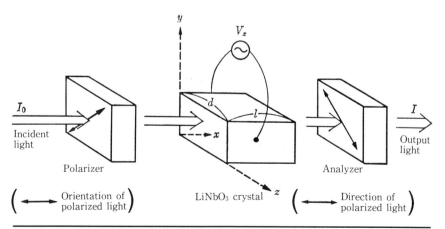

Figure 10.12. *Example of light modulation.*

where λ is the wavelength of the light. The output intensity of the light beam is $I = I_0 \sin^2(\Gamma/2)$. Here, if the crystal plate with Γ as $\pi/2$ or π has been set, the crystal plate is called the $\lambda/4$ or $\lambda/2$ wave plate and its output is $I_0/2$ or zero.

Now there are a new phase difference Γ'_x and a new output light beam intensity I' when the voltage $V_z = E_z^0 d$ is applied over the z surfaces with distance d. If the change in the refractive index due to E_z^0 is

$$1/n'_0 = (1/n_0)^2 + \gamma_{13} E_z^0,$$
$$1/n'_0 = (1/n_e)^2 + \gamma_{33} E_z^0,$$

then, approximately,

$$n'_0 = n_0 - (n_0^3 \gamma_{13} E_z^0)/2,$$
$$n'_e = n_e - (n_0^3 \gamma_{35} E_z^0)/2.$$

Thus, we can obtain

$$\Gamma' \equiv \Gamma + \Gamma'' = \Gamma + \frac{2\pi l n_e^3 \gamma_c V_z}{2\lambda d}$$

and

$$\gamma_c \equiv \gamma_{33} - (n_0/n_e)^3 \gamma_{13}.$$

Of them, Γ is invariant with respect to the applied voltage and corresponds to the bias light component in the optical modulation

$$I' = I_0 \sin^2 (\Gamma/2 + \Gamma''/2).$$

Now if with the adjustment so that $\Gamma = \pi/2$, then $I' = I_0 (1 + \sin^2 \Gamma'')/2$, and, when V_z is small, is proportional to V_z and varies in the form

$$I' \doteq (1 + \Gamma'')/2.$$

This result is one example of optical modulation by the Pockels effect.

Let us focus on the phase δ'_x and δ'_y of $E_x^{\omega'}$ and $E_y^{\omega'}$ at the output point of an LN crystal V_z changes with the angular frequency ω_0 ($V_z = V_{z0} \sin \omega_0 t$). Thinking of them in the same way as the phase difference Γ', we find that in

$$\delta' x \equiv \delta_x + \delta''_x = (2\pi n'_0 l)/\lambda$$

and

$$\delta'y \equiv \delta y + \delta''y = (2\pi n'_e l)/\lambda$$

δ''_x and δ''_y are proportional to V_z. Therefore, since $\delta''_x = \delta_{x_0} \sin \omega_0 t$,

$$E_x^{\omega'} = E_{x0}^{\omega} \exp[j(\omega t + \delta_x + \delta_{x_0} \sin \omega_0 t)]$$

and, by nonlinearity, waves of frequency $\omega \pm m\omega$ (where m is an integer) are included in $E_x^{\omega'}$ ad $E_y^{\omega'}$. Now, if ω_0 is the frequency of light beam ω and m is 1, then the second harmonic generation is found. This result indicates that γ and **d** are basically in correspondence. Note that, in actual second harmonic generation ω and 2ω will have different refractive indices due to the dispersion. The electrical field of incident light beam E^ω propagates at a phase velocity of c/n^ω, exciting polarization $P^{2\omega}$ in the crystal. The electrical field of light $E^{2\omega}$ which occurs due to $P^{2\omega}$ propagates at a phase velocity of $c/n^{2\omega}$, giving an $E^{2\omega}$ intensity of

$$I^{2\omega} \propto (n^\omega - n^{2\omega})^{-2} \sin^2[2\pi l(n^\omega - n^{2\omega})/\lambda]$$

at the crystal output end. Since $I^{2\omega}$ takes on its maximum value at $n^\omega = n^{2\omega}$, the appropriate choice of crystal orientation and crystal temperature realizes the maximum $I^{2\omega}$. It is advisable to make the refractive index correspond with ordinary dispersion $n^{2\omega} > n^\omega$ and the uniaxial negative crystal $n_0 > n_e$ to $n^{2\omega} \to n_e$ and $n^\omega \to n_0$. For example, for **d**$_{31}$ of LN, if a 1.06 μm YAG laser light is transmitted in a direction about 84° from the z axis, an intense green light (0.53 μm) is obtained.

This discussion, including the two applications roughly sketched, is only the barest beginning of an outline of the optical properties of crystals. There has been heightened interest in their optical characteristics since the discovery of lasers. Before then, however, they had also been used widely in many fields. Here we cannot touch on all the possible topics and details. For those wishing to pursue the subject further, Ogawa's (1980) and Mason's (1966) excellent books on the subject are recommended.

REFERENCES

Ariyama, Kanetaka, et al. (eds.). (1958). *Bussei butsurigaku kōza* [The physics of physical properties]. 9 Vols. Tokyo: Kyōritsu Shuppan.

Geusic, J. E., et al. (1964). Laser oscillation in Nd doped yttrium aluminum, yttrium gallium and gadolinium garnets. *Applied Physics Letters,* 4, 182–184.

Jona, Franco and Shirane, Gen. (1962). *Ferroelectric Crystals*. New York: Pergamon.

Kesshō kōgaku handobukku henshū iinkai. (ed.) (1971). *Kesshō handobukku* [Handbook of crystals]. Tokyo: Kyoritsu Shuppun.

Koide, Shigenao, et al. (1969). Growth and some characteristics of single crystals for optoelectronic use. *Toshiba rebyū* [Toshiba review], 24, 1516–1525.

Maiman, T. H. (1960). Stimulated optical radiation in ruby. *Nature (London)*, 187, 493–494.

Mason, W. P. (1966). *Crystal physics of interaction processes.* New York: Academic Press.

Mitsui, Toshio, et al. (ed.) (1969). *Kyō yūdentai* [Ferroelectrics]. Tokyo: Maki Shoten.

Mitsui, Toshio, et al. (ed.) (1969). *Kyō yūdentai* [Ferroelectrics]. Tokyo: Maki *Relationships in Science and Technology*. New Series, Group III. Vol. 3. Berlin: Springer-Verlag.

Nye, J. F. (1960). *Physical Properties of Crystals.* Oxford: The Clarendon Press.

Ogawa, Norisuke. (1980). *Kesshō butsuri kōgaku* [Crystal physics engineering]. Tokyo: Shōkabō.

Pamplin, B. R. *Crystal Growth.* (1975). New York: Pergamon.

Shimoda, Kōichi et al. (eds.) (1972). *Ryōshi erekutoronikusu* [Quantum electronics]. Vol. 1. Tokyo: Shōkabō.

Takei, Fumihiko, et al. (1969). Growth of single crystals for laser application. *Toshiba rebyū* [Toshiba review], 24, 1507–1515.

11
Biological Applications

Kazuo Inamori
Kyocera Corporation
Yamashina-ku, Kyoto 607, Japan

I. METALLIC IMPLANTS

Most bone and joint repairs attempted in humans over the centuries have relied on metal devices because of their rigidity, malleability, and strength. (See Table 11.1.) Gold was apparently first identified for use in the human body as bone plates, as long ago as A.D. 1500–1600. Sherman's vanadium-steel plates were used around 1910–1920, but even then, corrosion and low strength were easily identified in this early application as two major problems facing metal implants. There have been many developments since then using high-strength and more corrosion-resistant alloys as load-bearing structures for dental and orthopedic implants. However, only three classes of metallic alloys have been deemed biomedically acceptable for structural implants, that is, stainless steel (iron-based), cobalt-based, and titanium-based, that is, the Ti-6Al-4V alloy.

As a direct result of this considerable experience, the metallic systems remain with us today, even though it is recognized that they are not as innocuous as they were once thought to be. It is now well known that metals corrode and release ions or particulate material into the tissues. The concentration of these metallic ions in the body is in direct proportion to the surface area of the metallic implant. Even the titanium-6Al-4V alloy, which has the best corrosion resistance and compatibility of any of the

Table 11.1. History of Metallic Implants

Alloy	Device[a]	Year
Gold	Bone plates	1500–1600
Iron	Fixation wire	1700–1800
Vanadium steel	Bone plates	1910–1920
Cobalt chrome, cast	Joint replacements	1938–present
Stainless steel 302	Varied implants	1938
Stainless steel 18-8 SMo	Varied implants	1946
Titanium, commercially pure	Varied implants	1953
Ti-6Al-4V alloy	Joint replacements	1957
Stainless steel 316L, annealed	THR stem	1961
Stainless steel 316LVM, annealed	THR stem	1966
Stainless 316LVM, cold-worked	THR stem	1971
Stainless steel, cast	THR stem	1971
Cobalt nickel chrome, forged	THR stem	1972
Cobalt chrome, isostatically hot-pressed	THR stem	1975
Cobalt chrome, forged	THR stem	1978
Stainless steel, N_2 enhanced	Fracture fixation	1981

[a] THR, total hip replacement.

metals, increases the body burden of the metal ions titanium, aluminium, and vanadium. The body's response is a reaction directly related to the rate of release of those products and their chemical nature such as toxicity. This response can be local or systemic at the bacteriological, toxicological, carcinogenic, immunologic, and metabolic levels. The contaminants can be released as a result of diffusion of metal ions through the surface passive layer, from corrosion generally, and from wear.

In terms of biocompatibility, Laing et al. (1967) noted that iron, nickel, chromium, cobalt, manganese, molybdenum, and vanadium produced undesirable cellularity and fibrosis. As a result of these studies, they suggested that it would be preferable to avoid these elements in medical implants.

A variety of pathologies have been identified in relation to the various elements in such metallic implants, varying from dermatologic reactions to local tissue necrosis and tumors. Recently there has been concern about the extended long-term use of metallic joint replacement systems in young patients. Hamblen and Carter (1984) stated that, with regard to tumor formation, "Surgeons performing metal joint replacements with non-cemented prostheses, particularly in younger patients, must enlarge their awareness of this remote but potential risk."

The question is how to replace the metallic systems with more biocompatible materials with improved functional capabilities.

II. BIOLOGICAL PRACTICE IN OPTIMIZING IMPLANTS

The three fundamental points now facing implant manufacture and technology are:

1. To obtain biocompatible materials which are tolerated in the biological adverse environment for many years of use
2. To create designs which guarantee reasonable mechanical stability and strength
3. To settle the implant in its bony bed for long-term fixation even in young patients

The prerequisite for any biomaterial to be implanted in the body is that it not injure the biological system nor be damaged by it. That is, this material should not release any constituents into the body that will injure or kill the cells and it should be impervious to attack by the body's defense mechanisms. Thus the ideal material should be physically and chemically neutral. Obviously the metallic implants have not met this criterion.

One class of nonmetallic materials which excel in biocompatibility and inertness is the ceramic group. The driving force behind ceramic developments in biomedical applications stems from the fact that we have become sophisticated in our understanding of how the body accepts or rejects artificial implants. Materials are required that will function long term in intimate apposition with either soft tissues or bone.

III. CARBON AND CERAMIC SYSTEMS

The field of ceramic materials is vast, with many variations; therefore researchers have made their own sets of optimal properties and form. A general definition of ceramic materials is "those which have as their essential component, and are composed in large part of, inorganic materials" (Kingery et al., 1976). Ionic bonding is the predominant atomic feature of ceramics, which are, by definition, structural solids rather than nonmetallic elements. This implies that ceramics always contain more than one kind of element.

There are many classes of ceramics with unique attributes for medical implants, including crystalline ceramics, glasses, glass-ceramics, and carbons. Ceramics either under clinical use or development generally can be classed as one of three types (see Table 11.2):

1. Bioinert (nonactive, nondegradable)

Table 11.2. Types and Strength of Medical Ceramics

Type	Flexural strength (MPa)
Bioinert ceramics	
Alumina, single-crystal	1,300
Alumina, polycrystalline	380
Vitreous carbon	225
Pyrolitic carbon (LTI)	520
Surface-active ceramics	
Calcium phosphate glasses	110–150
Hydroxy apatite	110–190
Biodegradable ceramics (TCP)	
Fresh human bone (compact)	170

2. Controlled surface activity
3. Totally resorbable

A. CRYSTALLINE AND VITREOUS CARBONS

Carbon is either crystalline or vitreous. The graphite and diamond types of carbon are the two main types of crystalline arrangements of carbon atoms. The graphite type has relatively strong bonding between the atomic hexagon structures in the graphite layers but weak bonding between the layers. The low-temperature isotropic carbon layers are deposited at elevated temperatures (1300°–1500°C) from pure hydrocarbon gas onto a preformed refractory substrate, for example, graphite. The low-temperature isotropic carbon microstructure makes crack propagation extremely difficult, requiring a large amount of energy to proceed (Bokros et al, 1975). Even greater strength can be obtained for low-temperature isotropic carbons by alloying with silicone.

Vitreous carbon is a glassy form of high-purity carbon (99.9%), formed by pyrolyzing phenol resin (1800°–2000°C), driving off the volatile constituents, and leaving a glassy-appearing carbonaceous residue.

B. BIOINERT OXIDE CERAMICS

Alumina has been developed extensively as a structural ceramic. Scientific investigations of alumina go back to the last century but the first identification of commercial use was with the introduction of a patent for production of alumina (in 1907).

The crystal structure of the oxide ceramics is typified by the closest packing of the oxygen ions with the metal ions located in the interstices.

For example, alumina (corundum ceramic) has a hexagonal closest packed arrangement of oxygen ions, the six oxygen ions completely surrounding the aluminum ion. This corundum or alpha alumina phase is crystallographically identical to sapphire and ruby, the alpha phase being the most stable modification of alumina. Thus, alpha alumina represents one of the most stable of the inert ceramics. Single-crystal implants are grown from one pure crystal of ceramic and then machined to the final shape, whereas the polycrystalline product is composed of many millions of ceramic particles fused together after some shaping in the green stage. High-purity alumina ceramic has also played an important role as a bone substitute for orthopedic and dental implants, as well as in the field of maxillary and oral surgery. (See Figures 11.1 and 11.2).

C. BIODEGRADABLE CERAMICS

Since the constituents of a resorbable implant must be metabolized as the implant dissolves, the composition of such ceramics is extremely critical. Acceptable formulations basically involve CaO and P_2O_5 compounds, such as $Ca_3(PO_4)_2$.

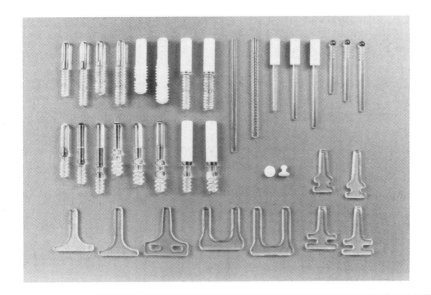

Figure 11.1. Artificial dental roots of Bioceram single-crystal alumina.

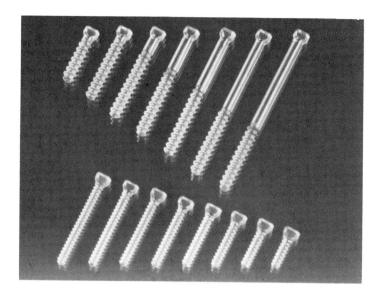

Figure 11.2. Bone screws of Bioceram single-crystal alumina.

D. SURFACE-ACTIVE CERAMICS

The concept behind surface-active ceramics is to provide a ceramic surface layer on an implant which will enhance the implant fixation by virtue of a controlled chemical reaction between the ceramic and the surrounding bone tissues.

The available data indicate that the strength and stability of the interfacial bond between bone and bioglass are due to chemical bonding of the organic-inorganic constituents in the intercellular transition layer (Hench, 1981).

In contrast, bioactive ceramics, such as dense hydroxyapatite, appear to form a bond with the interfacial core and by the growth of apatite (Hench, 1985).

IV. MEDICAL APPLICATIONS

A. MATERIALS

1. Carbon

Carbon has been studied for medical devices since the late 1960s (Bokros et al., 1970). Carbon has been used extensively in cardiovascular devices because of its excellent thromboresistance (Bokros et al., 1972). (See

Figure 11.3). Other problems that the use of carbon could solve include transcutaneous devices for attachment of artificial limbs, grommets for taking electrical leads through the skin, dental implants (Kojima et al., 1984), repair of cartilage defects (Minns et al., 1972) and transducers to improve the hearing mechanism. Recent developments have centered around the deposition of a thin, adherent, impermeable layer of pure isotropic carbon on the surfaces of high-strength metal implants or even polymeric components, for example, heart-valve components and artificial hip joints.

Carbon fiber has also been found useful owing to its great strength, reinforcing, ability, and pliability. It has been used in bone cements (Pal and Saha, 1983), polyethylene composite ankle joints (Groth and Shilling, 1983), and ligament replacements (Weiss et al., 1985).

Carbon has also been applied to orthopedic devices, including high-strength isotropic carbon, silicon carbide/carbide composites, and the carbon-fiber-reinforced matrix of either carbon or other media such as polymethyl-methacrylate or polysulfone, for example, bone plate and hip cup (Roberts and Ling, 1980) and the fixation layer on the total hip.

The stiffness of this material is about the same that of cortical bone, that is, much less than metal or ceramic materials. These properties have made

Figure 11.3. Artificial heart valve of low-temperature isotropic carbon.

low-temperature isotropic carbon the material of choice for long-term implantation in heart valves and for use in cardiac-assist devices. Current specifications for heart valves include up to 15% silicone (Bokros et al., 1975). Fatigue studies have shown that up to 10% silicone can raise the endurance limit from the 340 to 550 MPa range, that is, 60% improvement (Shim, 1974). Low-temperature isotropic carbons also have excellent wear performance and are not subject to biodegradation.

2. Alumina

Most ceramics used for implantable devices belong to the group of oxide ceramics. Alumina has been the most intensively developed implant ceramic. The alumina used for medical and dental implants is a highly pure material which comes in either single-crystal or polycrystalline (ASTM-F603) forms. The strength of alumina has proved to be excellent for many implant applications. It is also the most rigid of the current implant materials.

Compared to metals, ceramic mechanical properties are characterized by:

1. Low tensile strength
2. Dispersion of strength values
3. Brittleness
4. Susceptibility to mechanical and thermal impact loading

Ceramic materials have a unique combination of desirable properties:

1. Chemical inertness
2. Excellent biocompatibility
3. High corrosion resistance
4. High wear resistance
5. High compressive strength
6. High creep resistance
7. High hardness

The amount of aluminum lost in the body from the major part of the surface must be negligible in pure dense alumina. Only a very thin hydrated layer $Al(OH)_3$ is formed on the surface in contact with body fluid, less than 50 Å thickness (Williams, 1981). Alumina has a hydrophilic surface layer which results in the coating of its surface with a layer of water molecules. This adsorbed layer of water and other material provides the ceramic with an excellent camouflage and aids in its acceptance by the body.

3. Multicomponent

There is considerable interest in those ceramics which appear similar to the mineral phase of bone. These include hydroxyapatite, $Ca_5(PO_4)_3OH$; octacalciumphosphate, $Ca_4H(PO_4)_3 \cdot 2H_2O$; and tricalcium phosphate, $Ca_3(PO_4)_2$, commonly referred to as TCP.

B. CLINICAL APPLICATIONS

Current medical usage of ceramic materials includes artificial joints and teeth, partial or total replacements of bones, fillers for bone defects, reconstruction of cosmetic defects and even usage in selected components of artificial organs, for example, valves for artificial hearts, and so on. Other biomedical applications include various sensors in gas and temperature monitoring equipments. The particular application discussed in this review in the use of alumina ceramic for bone and joint replacement.

The exciting possibilities of using alumina ceramics in dental and orthopedic applications was foreseen by Kawahara in 1965. The ceramic which appeared to meet the need for a high-strength, inert material for loaded structural implants was aluminum oxide. Such strength requirements can be met by high density alumina with a small grain size.

After in vitro and animal studies, the stage was set for the first clinical study of the alumina single-crystal screw dental implant (Kawahara et al., 1975, 1980). These studies verified the predictions of the research program and single-crystal implant procedures continue worldwide. In contrast, the Mitre company in the United States continued the development of the weaker polycrystalline implant (Driskell and Heller, 1977).

In one of the most extensive and detailed follow-up studies of single-crystal implants, McKinney and colleagues analyzed single-crystal screw endosteal dental implants in animals clinically, histologically, ultrastructurally, and statistically (McKinney et al., 1982, 1983). Their study illustrated the excellent biocompatibility and functional coadaptation of bone and soft tissues to the single-crystal dental implant.

C. ARTIFICIAL JOINTS

Alumina ceramic gives optimal tribological characteristics to the joint replacement. This results in a specially adapted surface layer of high bone strength which provides the alumina grinding surface with very low friction and optimal wear resistance. (See Figure 11.4).

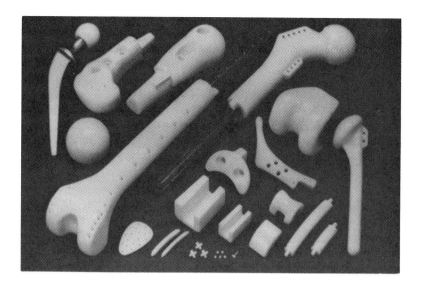

Figure 11.4. Artificial alumina bones and joints (Bioceram).

The human hip joint is the most commonly replaced joint when it becomes arthritic. Current estimates indicate that about 200,000 hip joints are replaced each year all over the world. Alumina ceramic is the optimal choice for the bearing of the femoral ball because of its excellent low-friction properties. As a material sliding on UHMWPE (ultrahigh molecular weight polyethlene), it can decrease the friction associated with a metal–UHMWPE bearing by a factor of four. The combination of ceramic–UHMWPE materials is the superior choice. (See Figure 11.5).

Another very important topic today, especially for maximizing long-term survival in younger patients, is the reduction of wear debris invading the joint tissues. Ceramics also offer major improvements here. Laboratory hip simulator studies have demonstrated improvements in wear resistance for ceramic–UHMWPE total hip replacements, better than metal ball–UHMWPE cups by 2 to 16 times.

Therefore total hip replacement with the ceramic ball and UHMWPE cups offers considerable advantages for younger patients:

1. No insult from ceramic particles (no ball wear occurs, in direct contrast to ceramic–ceramic total hip replacements)
2. No toxic ions such as cobalt, nickel, chromium, and so on
3. At least three times less insult to the joint tissues, as a result of the much reduced volume of UHMWPE wear particles

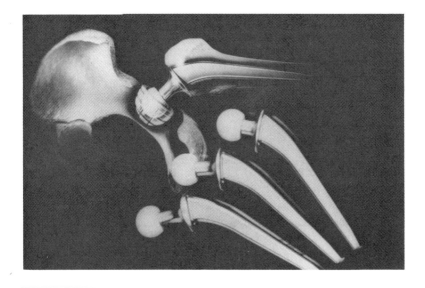

Figure 11.5. THP: a combination of alumina head, metal stem, and HDP socket (Bioceram).

4. Longer service with optimal range of motion as a result of less medial migration of the femoral ball.

Because many people do not understand that ceramic materials are brittle and therefore dangerously weak for implants, the manufacturers conduct many mechanical studies of each implant system. These may include studies of friction, wear, impact strength, compressive strength, fatigue strength, and so on. Generally these tests are conducted using severe loads. Accordingly the implant design and mechanical research and development test programs are of paramount importance to obtain the optimal combination of material and device properties.

We note that the weight ranges of European and American patients are typically higher than those of Asian patients. Therefore the loads on the prostheses can be expected to be correspondingly higher. In addition, owing to the biomechanical effects of gravity and muscles on the joints, the actual implant loads are several times higher than the patient's actual weight. The ceramic hip implant must be carefully designed and tested to ensure that it can withstand such loads for many years in the human body.

Given that the ceramic implant passes such severe tests with a very large safety factor, it seems reasonable to assume that even small variations in fabrication properties and variations in patient weight and activity levels will be safe over many years of usage in the human body.

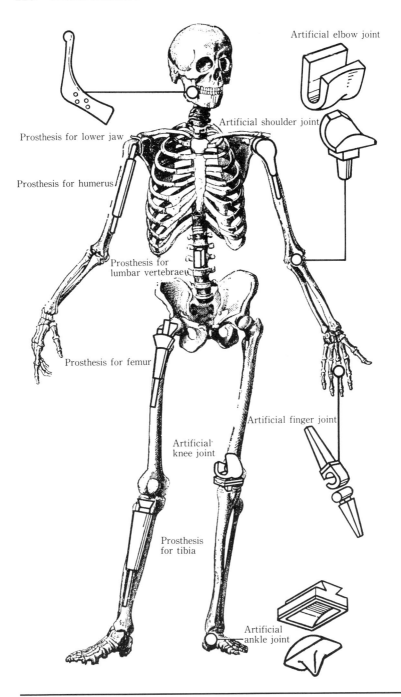

Figure 11.6. Examples of applications of Bioceram artificial bones and joints.

V. CONCLUSION

Recently ceramics have been used regularly as implant materials in the field of medicine and dentistry. (See Figure 11.6). Among them, alumina ceramics especially have given excellent results, and it is expected that, in the near future the application range of single-crystal and polycrystalline aluminas and their combinations will be enlarged, and that using such ceramics with other materials, along with the new ceramics, to simulate the characteristics of human hard tissue will continue. For future development, research must be conducted in the fields of physics, engineering, and so on, not just in the field of medicine.

REFERENCES

Bokros, J. C.; Gott, V. L.; LaGrange, L. D.; Fudali, A. M.; and Ramos, M. D. (1970). *Journal of Biomedical Materials Research,* 4, 145.
Bokros, J. C.; LaGrange, L. D.; and Schoen, F. J. (1972). In P. L. Walker (ed.), *Chemistry and Physics of Carbon.* New York: Marcel Dekker, pp. 103–171.
Bokros, J. C.; Akins, R. J.; Shim, H. S.; Haubold, A.; and Agarwal, N. K. (1975). *American Chemical Society Symposium No. 21.,* p. 237.
Driskell, T. D. and Heller, A. L. (1977). *Journal of Oral Implants,* Vol. 7, no. 1, 53–76.
Groth, H. E. and Shilling, J. M. (1983) *Journal of Orthopedic Research,* 1, 129–135.
Hamblen, D. L. and Carter R. L. (1984). *Journal of Bone and Joint Surgery,* 66B, 625–627.
Hench, L. L. (1981). In D. F. Williams (ed.), *Fundamental Aspects of Biocompatibility.* Vol. 1. Boca Raton, Fl: CRC Press, pp. 67–85.
Hench, L. L. (1985). In *11th Ann. Meet. Soc. Biomat.,* San Diego, April 25–28, p. 25.
Kawahara, H. (1965). *Second Proceedings of the International Academy of Oral Pathology,* September 1–4, Melbourne, Australia. pp. 79–91.
Kawahara, H.; Yamagami, A.; and Hirabayashi, M. (1975). *The First Proceedings of the Japan Society of Implant Dentistry,* pp. 187–196.
Kawahara, H.; Hirabayashi, M.; and Shikita, T. (1980). *Journal of Biomedical Materials Research,* 14, 597–605.
Kingery, W. D.; Bowen H. K.; and Uhlmann, D. R. (eds.) (1976). *Introduction to Ceramics.* New York: John Wiley and Sons.
Kojima, A.; Otani, S.; Yanagisawa, S.; Sairenji, E.; and Nijima, K. (1984). *Carbon,* 22, 47–52.
Laing, P. G.; Ferguson, Jr., A. B.; and Hodge, E. S. (1967). *Journal of Biomedical Materials Research,* 1, 135–149.
McKinney, R. V.; Koth, D. L.; and Steflik, D. E. (1982). *Journal of Oral Implantology.,* 10, 487–503.
McKinney, R. V.; Steflik, D. E.; and Koth, D. L. (1983). *Journal of Oral Implantology,* Vol. 11 no. 3.

Minns, R. J. (1987). *J. Med. Eng. & Tech.*, 7, 200–201.
Pal, S. and Saha, S. (1983). In *Trans. Biomat. Soc. 1983.* p. 35.
Roberts, J. C. and Ling, F. F. (1980). J. Mech. Des., 102, 688 694.
Shim, H. S. (1974). *Biomat. Med. Dev. Artif. Org.*, 2, 55–65.
Steflik, D. E.; McKinney, Jr., R. V.; and Koth, D. L. (1983). *J. Dent. Res.*, 62. 1212–1215.
Weiss, A. B.; Blaziana, M. E.; Goldstein, A. R.; and Alexander, H. (1985). *Clin. Orthop.*, 196, 77–85.
Williams, D. F. (1981). *Fundamental Aspects of Biocompatibility*. Vol 1. Boca Raton, Fl: CRC Press, 1981.

12

Mechanical Properties

Shigetomo Nunomura
Tokyo Institute of Technology
Midori-ku, Yokohama 227, Japan

Junn Nakayama and Hiroshi Abe
Asahi Glass Co., Ltd.
Kanagawa-ku, Yokohama 221, Japan

Osami Kamigaito
Toyota Central Research & Development
Laboratories, Inc.
Aichi-gun, Aichi 480-11, Japan

Kitao Takahara and Katsutoshi Matsusue
National Aerospace Laboratory
Chofu, Tokyo 182, Japan

I. STRENGTH OF MATERIALS

Interest in the mechanical strength of ceramics is increasing, to give full scope to their functional characteristics. Ceramics are also receiving much

Note: Authorship of this chapter is as follows: section I is by Shigetomo Nunomura, section II by Junn Nakayama and Hiroshi Abe, section III by Osami Kamigaito, and section IV by Kitao Takahara and Katsutoshi Matsusue.

attention as load-bearing structural materials with heat resistance or biological compatibility.

The concept of strength is the extent to which the various expectations of users are fulfilled. A tacit understanding on strength-related standards has been reached for existing structural materials, but this understanding cannot be applied as-is to the new materials.

The strengths looked for by users can be classified according to use: (1) minimal deformation under load, (2) high load before permanent deformation begins, and (3) high load and energy required before fracture. For metals, these correspond, respectively, to the coefficient of elasticity, the yield point, and the fracture toughness. In metals, the fracture load is normally greater than the load required for deformation; thus, a metallic structural material cannot perform beyond its yield point, (2), and the fracture toughness, (3), rarely comes directly to the fore. In ceramics, however, as discussed below, the load required for fracture can be less than the load needed for deformation; therefore, a ceramic material's strength is expressed in terms of breaking strength only. This difference can cause confusion if the strength of a ceramic material is treated in the same sense as the yield point of metal materials.

A. DEFORMATION OF CERAMICS

For the ceramics which are the subject of this book, deformation means elastic deformation only at temperatures of 1000°C or below. Strength of materials gives Eq. (12.1) for the strength of an elastic material:

$$\varepsilon = \sigma/E, \qquad \gamma = \tau/\mu \tag{12.1}$$

Here, ε and γ are the tensile strain and shear strain, σ and τ are the tensile stress and the shear stress, E is Young's modulus, and μ is the rigidity. From a practical standpoint, Young's modulus is frequently measured, but measurement of the rigidity or of Poisson's ratio is rare. Table 12.1 gives values of Young's modulus, Poisson's ratio, hardness, fracture toughness, and deformation rate under high temperatures for representative substances. These values, however, are but a single example of measurement results; other values have resulted from measuring other samples.

Young's modulus for composite materials composed entirely of ceramic substances or of ceramics and other substances has been largely established to follow the rule of mixture given in Eq. (12.2).

$$E = E_a V_a + E_b V_b \tag{12.2}$$

V_a and V_b are the volume fractions. Young's modulus has been frequently

Table 12.1. Examples of the Fracture Strengths and Coefficients of Elasticity for Ceramics

Material	Young's modulus (GPa)	Poisson's ratio	Hardness	K_{Ic} (MPa·m)	High-temperature strength (10^{-5}cm/cm·hr) $\left(\dfrac{1,300°C}{12.5\ \text{MPa}}\right)^a$
Concrete	14	0.2	—	—	
NaCl	44	—	24	0.5	
Aluminum alloy	69	0.5	390	33	—
Fused silica	69	—	540–620	0.7	
Glass	69	—	—	—	
ZrO_2	138	—	1,000–1,400	7.6	10,000
Iron	200	0.3	200	200	3
MgO	207	0.36	620–920	1.2	—
Ni-based alloy	210	0.33	—	74	3.3–3.5
BeO	311	0.34	1,140	—	—
Al_2O_3	380	0.26	1,200–2,000	2.0–4.6	30
SiC	414	0.14	3,300	—	0.01–0.18
Diamond	1,035	—	9,000	—	—

[a] Creep at 1300°C under 12.5 MPa.

measured at high temperatures (Wachtman, 1955; Kawashima and Suzuki, 1964; Richerd, 1982). The following equation applies to alumina at a temperature range up to 600°C.

$$E = E_0 - bT \exp(-T_0/T)$$

Generally, E decreases with a rise in temperature; but it has also been found that E reincreases with another rise in temperature (Kawashima and Suzuki, 1964, p. 101). Porosity affects the coefficient of elasticity in the same way. The relationship in the following formula can be found to apply to a porosity ϕ of up to 50%, with a Poisson's ratio of 0.3 (McKenzie, 1950, p. 2).

$$E = E(1 - 1.9\phi + 0.9\phi^2)$$

Spriggs (1961) has carried out a wide range of experiments concerning the influence of porosity on mechanical characteristics. The deformation of ceramics, special single crystals excepted, at 1000°C and below is entirely elastic deformation; thus, maximum deformation is readily determined from the fracture load through Eq. (12.1). Plastic processing is not possible after sintering, but creep occurs at temperatures beyond 1000°C. There are numerous studies concerning this phenomenon (Kingery Bowen, and Uhlmann, 1976).

B. DEFORMATION STRENGTH AND BREAKING STRENGTH

In almost all ceramics, there is no deformation strength equivalent to the yield point of metals. Breakage, which occurs when the bonds between atoms fracture along a plane and split apart on a plane perpendicular to the tensile stress at that point, governs strength. Take a case in which a tensile or compression load P is imparted to the columns shown in Figure 12.1 in the direction of the axis. Let us consider the splitting of uniform pairs of closely linked atoms above and below the plane A–A', under tension, or on the plane C–C', under compression. Under compression, with an operative force which is ν times the tension (where ν = the Poisson ratio 0.2–0.5), the apparent strength is therefore greater than when the column is under tension.

First, let us deal via the roughest approximation with separation along the plane A–A'. Between two atoms, both universal gravitation and the repulsion based on Pauli's principle of exclusion are at work. These forces as a whole assume the form shown in Figure 12.2. Accordingly, for the two

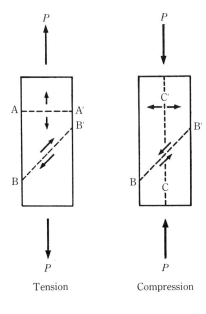

Figure 12.1. Slip planes A–A' and C–C', and slip plane B–B under tension and compression loads.

atoms to be separated, work is required commensurate with the area beneath that curve. This work must at least be equal to the increase in surface energy through exposure of the new plane. Approximating the tensile strength per unit area σ closely with Eq. (12.3),

$$\sigma = \sigma_F \sin(2\pi x/\lambda) \tag{12.3}$$

the quantity of work U is

$$U = \int_0^{\lambda/2} \sigma_F \sin(2\pi x/\lambda)\, dx$$

On the assumption that this work is equivalent to the surface energy, then Eq. (12.4) follows:

$$\lambda \sigma_F/\pi = 2\gamma \tag{12.4}$$

Near $x = 0$, however, Hooke's law applies. Thus, if b is taken as the distance between atoms,

$$\sigma = Ex/b = 2\pi x/\lambda$$

and when λ is eliminated using this equation and Eq. (12.4), Eq. (12.5)

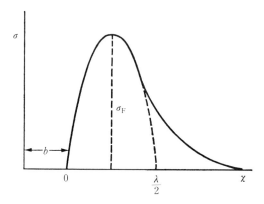

Figure 12.2. Force acting upon pairs of atoms and the work required to separate them.

obtains.

$$\sigma = (\gamma E/b)^{1/2} \tag{12.5}$$

The shearing force τ that gives rise to slippage along the plane B–B' is the same under both compression and tension. For crystalline substances, it can be derived in the following way. The crystal lattice points have their minimum potential energy U at stable locations of the atoms. These locations exist periodically at distances b. Since U can be expected to change smoothly in the interval of transition by slippage to the next stable location, the changes in U are approximated by a sine function, as shown in Figure 12.3.

$$U = U_0 \sin^2(\pi x/b)$$
$$\tau = \partial U/\partial x = (\pi U_0/b) \sin(2\pi x/b)$$

Near partial derivative $x = 0$, since Hooke's law applies, if d is taken as the interatomic distance between slip planes, then

$$\tau = \mu x/d$$
$$\therefore \quad \tau = (b\mu/2\pi d) \sin(2\pi x/b)$$

When $d = b$, $\tau_{max} = \mu/2\pi$. Moreover, considering the three-dimensional equilibrium position and calculating the stress in the direction of tension, we obtain Eq. (12.6).

$$\sigma_{max} = \sqrt{2}\mu/10\pi \tag{12.6}$$

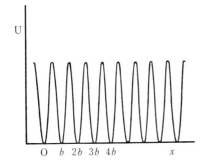

Figure 12.3. Position of atoms along a slip plane and their potential energy.

Table 12.2 presents values for breaking strength and yield point arrived at with Eqs. (12.5) and (12.6), together with actual measured values, for a selection of materials. The fracture load arrived at by calculation is larger than the deformation load; thus, before separation occurs along the planes A–A' and C–C', slippage occurs along the plane B–B'. That causes the structural material to lose its usefulness without reaching the breaking point.

As can be seen in Table 12.2, the actually measured strengths are much lower than the calculated values. It can be understood conceptually that the breaking strength is reduced due to the concentration of stress at acute latent defects, and that the strength will also be reduced as slippages occur at movements of line defects (dislocations). With almost all metals, the actual slippage strength is less than the breaking strength, so that major plastic deformation precedes fracture.

In ceramics, however, since the likelihood of acute latent defects is high, and because the movements of dislocations are limited, breaking strength is reached before slippage strength. Accordingly, deformations in ceramics are elastic deformations only; plastic deformation does not occur.

Table 12.2. Theoretical and Actual Values for Deformation and Breaking Strength

Material	Calculated value		Actual value	
	Fracture Eq. (12.5)	Slipping Eq. (12.6)	Fracture	Yield point
Fe	10,000	3,100	276	—
Zn	3,300	1,360	2	1.5
Ni	9,000	3,100	≫8	8
Al_2O_3	38,000	6,000	400	≫400
SiC	44,000	6,800	700	≫700

Recently, dislocation movement in metal materials has been inhibited in an effort to achieve higher strength. As a result, the slippage value has approached the breaking value, and it has become necessary to appraise strength in terms of breaking strength in the case of materials with manifest faults (such as notches). This development has made evaluation of the strength of materials on the basis of slippage deformation inadequate, and a new fracture mechanics has been pioneered. Since metals have a slippage strength approaching their high breaking strength, this new fracture mechanics is applied only to cases where there actually are faults. In ceramics, the difference between the two values is either almost nonexistent or the slippage strength exceeds the breaking strength. Thus, active adoption of a fracture mechanics is essential in evaluating the strength of ceramics.

C. BENDING TESTS

To derive the coefficient of elasticity and the rupture in bending load in ceramics, it is usual to carry out three-point or four-point bending tests. In this case, shearing stress and bending moment are distributed as in Figure 12.4. Since the range of elasticity may be considered alone, if Mb is the bending moment, y is the distance from the neutral axis, and I is the moment of inertia of the cross section, the stress σ is

$$\sigma = Mb \cdot y / 2I$$

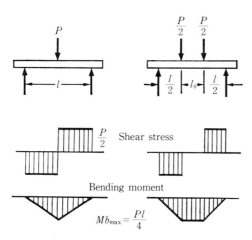

Figure 12.4. *Bending moment and shear stress in bending test samples.*

For a beam of width B and height b, $I = Bh^3/12$. Since the maximum surface stress is $Mb = Pl/4$ (see Figure 12.4), then

$$\sigma_{\max} = (3/2)(Pl/Bh^2)$$

The deflection, δ, is given by the equation below, and Young's modulus is derived from the slope of the δ–σ curve.

$$\delta = (1/6)(l^2/Eh)\sigma$$

The position of maximum stress is, in tensile tests, the entire volume of the test piece. In contrast, in three-point tests the position of maximum stress is on a straight line; with four-point tests, it is on a limited plane surface. The breaking strength depends on the existence of defects but, in bending tests, the probability that there will be major faults in the limited areas tested is low. For this reason, bending strength is higher than tensile strength and values are scattered more. In the case of slippage deformation also, the yield point according to bending tests is greater than values found in tensile tests.

However, with a dislocation density of $10^{10}/cm^2$ or above, this difference is not dependent on the distribution of defects, but follows from deformation detectability. Yield strengths evaluated by tensile tests and by bending tests are proportional to each other. Thus, the breaking strength according to bending tests is used in simple comparisons of the strength of ceramic materials and in quality control. For actual structural materials, however, it is impossible to give absolute strength values for design, since even if the objects will be subject to bending loads, their sizes differ and the distributions of defects vary.

D. FRACTURE MECHANICS TESTS

The intrinsic breaking strength of ceramic materials, not affected by the distribution of latent defects, can be derived by imparting sufficiently large artificial defects (notches) to test materials. Parameters of fracture dynamics include CTOD, the J integral, which takes into account plastic behavior, and the values G and K which treat fracture as linear elastic behavior. Since most ceramic materials can be treated as elastic, only the value K must be dealt with. (The value G is seldom used, and can be calculated from the value K.) The ASTM standard has been adopted internationally as the method of measurement for the linear fracture toughness value K_{IC} for metals (ASTM Standard Book). It would be desirable to calculate the fracture toughness of ceramics on the basis of this standard also. Figure 12.5 shows the form of bending test specimen to which this standard seems applicable for ceramics. In addition, there are CT and C-type test specimens that should be suitable.

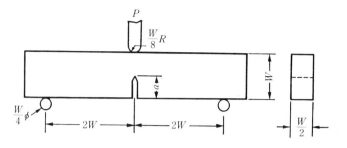

Figure 12.5. *Fracture toughness test specimen.*

Under this standard, to maintain consistency in the nature of the artificial defects, fatigue cracks are induced at the tip of the notch. It is thought that that step can be omitted in ceramics testing. Metal breaking strength is higher t' an slipping strength, and slight slipping deformation precedes fracture at the notch tip. To influence the measured value, fatigue cracks are essential. Ceramics, however, have lower slippage strength than breaking strength, and thus there is thought to be no need to induce fatigue cracks. There is, however, no experimental confirmation of this assumption. When the test specimen sketched in Figure 12.5 is used, the fracture toughness value K_{IC} is given by Eq. (127).

$$K_{IC} = (Pl/BW^{3/2})f(a/W)$$

$$f(a/W) = \frac{3(a/W)^{1/2}[1.99 - (a/W)(1 - a/W)(2.15 - 3.93a/W + 2.7a^2/W^2)]}{2(1 + 2a/W)(1 - a/W)^{3/2}}$$

(12.7)

P, a, W, and B are the fracture load and the notch length, width, and depth, respectively. The value K_{IC} derived is the intrinsic value for the material tested and is not influenced by the dimensions of the structural material or the test sample. The values for a sample of materials are given in Table 12.1.

The value K, as can be seen from Eq. (12.7), takes the form $K = \sigma\sqrt{\pi a}f(a/W)$, with $f(a/W)$ being determined by the dimensions of the sample or object tested. This value has been calculated for typical forms, and is summarized in the literature (Tada, Paris, and Irwin, 1973; Rooke and Cartwrite, 1974). By referring to these sources, the value of K_{IC} can be obtained for strength comparisons, and can be used to derive the maximum permissible load where the maximum defect dimensions are known or to derive the maximum permissible defect dimensions where the load is known.

II. MECHANICAL PROPERTIES OF TODAY'S CERAMICS

In recent years interest has risen in ceramics as a high-performance material. Valued as the "third material" after metals and plastics, and sometimes even labeled a dream high-performance material, ceramics have come to be referred to as *fine ceramics* and *engineering ceramics* by trade journals. These terms have even entered common usage through newspapers and television to become virtually everyday expressions today.

There are two basic reasons for the recognition that ceramics, especially those having special mechanical properties, have received: (1) the necessity for high-strength materials which can be used at high temperatures and (2) the potential that ceramics possess for further development. Both these factors have appeared only in the past 10 years.

Ceramics enjoy several major advantages over metals and other materials: their superior heat resistance, hardness, wear and abrasion resistance, corrosion resistance, and lightness. On the other hand, it is true that ceramics are inferior in terms of brittleness, especially with their low impact and thermal shock resistance. Those defects place major restrictions on applications.

These problems of brittleness have been overcome to a large extent with the development of silicon nitride and silicon carbide, as well as Sialon, all of which have excellent strength, high thermal conductivity, and relatively low thermal expansion coefficients. Moreover, zirconia, which has prospects as a material for adiabatic engines, is a ceramic with the potential to overcome the problem of brittleness. In this section, first the current state of development and usage of these materials and then the prospects for the future are discussed.

A. METALS AND CERAMICS AS MATERIALS WITH MECHANICAL CAPABILITIES

The comparison between the properties of metals and ceramics shown in Table 12.3 indicates the great differences between the two (Abe, 1982b, p. 62; Iseki, 1982, p. 605). The basic reason for the gap is the difference in the ways in which interatomic (and ionic) bonds form. In metallic bonds, electrons are not strongly bound by attraction to the nucleus, so that plastic deformation based on their dislocations can readily occur. As a result, metals have the advantage that fracture due to stress concentration is less likely to occur. In contrast, the ionic and covalent bonds in ceramics give the substance a structure in which atoms and ions cannot move easily. Thus, relaxation under stress is less likely to occur in ceramics; that characteristic has the disadvantage that breakage will result from comparatively small faults and that a ceramic object may even shatter suddenly

Table 12.3. Comparative Features of Metals and Ceramics[a]

Features	Metals		Ceramics[b]	
Heat resistance (°C)	1,100		1,300	
Corrosion resistance	X		O	
Strength features				
Toughness (MN·m$^{-3/2}$)	Low-strength carbon		Si_3N_4	5.3
	steel	>220	SiC	4.5
	Maraging steel	93	HP Al_2O_3	5.0
	Ti alloy	38	High toughness ZrO_2	10
	Al alloy	34		
Deformation when breakage				
occurs (%)	≥5		0.2–1	
Weibull coefficient	≥20		5–20	
Hardness (kg/mm^2)	Hundreds		2,000	
Fatigue characteristics	Plastic deformation		Crack growth	
Shock resistance (J/cm^2)	10–15		10^{-2}–10^{-1}	
Thermal shock resistance	Superior		Generally inferior	
Standardization of measuring				
method	Advanced		Lagging	
Previous record of mechanical				
capabilities	Extensive		Slight	

[a] The numbers in the table are rough values. From Abe (1982b, p. 62) and Iseki (1982, p. 605).
[b] HP, hot press.

without warning. Ceramics, however, do have the advantage of stability, with a high melting point (decomposition temperature), with no change in their properties even if exposed to chemicals, and with no change in shape even if external force is applied.

B. DISTINCTIVE FEATURES OF NONOXIDE HIGH-STRENGTH CERAMICS

Many of the engineering ceramics that have been receiving attention recently are nonoxides, for example, SiC, Si_3N_4, and Sialon (a solid solution of β-Si_3N_4 and Al_2O_3, expressed by the chemical formula $Si_{6-z}Al_zO_z$-N_{8-z}). One reason for emphasis on nonoxides is their great strength, compared with conventional ceramics (30–100 kg/mm^2 at room temperature), but perhaps even more significant is their low thermal expansion coefficients ($\alpha_{Si_3N_4} \approx 34 \times 10^{-7}$/°C, $\alpha_{SiC} \approx 43 \times 10^{-7}$/°C, $\alpha_{Al_2O_3} \approx 80 \times 10^{-7}$/°C). At the same time, because they are excellent conductors of heat (at around room temperature, $k_{SiC} \approx 50$ Kcal/m·hr·°C, $k_{Si_3N_4} \approx k_{Al_2O_3} \approx 20$ kcal/m·hr·°C), their resistance to thermal shock is far superior to that of conventional ceramics such as Al_2O_3.

Because the distinctive features of ceramics lie in these special high-temperature properties, they have potential for use in high-temperature ranges where metals cannot be employed. In such applications, of course, thermal shock resistance is a problem. In brittle materials such as ceramics, the higher the thermal shock fracture resistance coefficients R and R', the better the thermal shock resistance (Nakayama, 1977, p. 150). For this reason, it is a considerable advantage in making practical use possible in high-temperature situations to have high strength σ_C and high thermal conductivity k combined with a low thermal expansion coefficient α.

$R = \sigma_C(1 - \nu)/E\alpha$ applies in conditions of sudden thermal shock and $R' = Rk$ applies in conditions of gradual thermal shock. Here, ν represents Poisson's ratio and E Young's modulus.

The thermal shock frature resistance coefficients calculated on the basis of the physical properties of high-strength ceramics are shown in Figure 12.6 (Udagawa, 1982, p. 164). As the figure indicates, Si_3N_4, with its high strength and low coefficient of thermal expansion, is outstanding in fracture resistance under sudden thermal shock. Thermal conductivity is an important factor in resistance to gradual thermal shock. Thus, the thermal shock resistance of SiC, with its high thermal conductivity, is also superior.

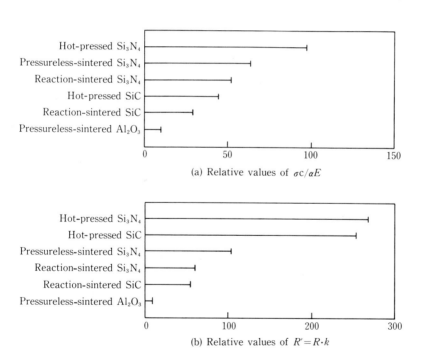

Figure 12.6. Thermal shock resistance of high-strength ceramics. (a) shows the coefficient of breaking resistance with respect to sudden thermal shock, and (b) shows the coefficient of breaking resistance with respect to gradual thermal shock. From Udagawa (1982, p. 164).

One of the outstanding features of ceramics, their strength at high temperatures, is shown in Figure 12.7. At relatively low temperatures Si_3N_4 is superior, while at high temperatures SiC is preferable. Metals are strong at around room temperature, but when the temperature rises, they suffer a rapid loss of strength (Yamada, 1982, p. 109). Apart from certain special alloys, metals cannot withstand temperatures above 1,000°C.

C. APPLICATIONS OF NONOXIDE HIGH-STRENGTH CERAMICS

It is obvious from the discussion of distinctive features so far that engineering ceramics such as SiC and Si_3N_4, which can maintain their strength at high temperatures, have a wide range of applications. The fields in which these ceramics can be utilized are listed in Table 12.4, while examples of products are shown in Figure 12.8 (Asahi Glass, 1982). The applications fields can be divided roughly into two types. One makes use of features of ceramics such as high elasticity, wear and abrasion resistance, and resistance to chemicals at relatively low temperature ranges. Such applications include the use of ceramics for mechanical seals, dies, gauges, bearings, cylinders, pistons, pumps, precision lathes, and various other types of machine tools. The other uses ceramics for machine parts at high temperatures, 1000°C and above. Overall, practical applications for lower tempera-

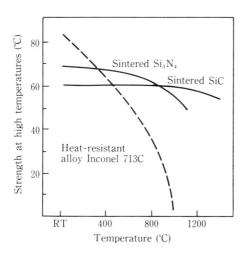

Figure 12.7. Comparison of the strength of ceramics and metals. Tensile strength measured for heat-resistant alloy; flexural strength for ceramics. (The tensile strength of ceramics is about half their flexural strength.)

Figure 12.8. Examples of nonoxide, high-strength ceramics.

tures are more advanced, while high-temperature applications are still largely in the development stages, aside from parts for the iron and steel industry.

At present, development projects are underway at individual enterprises, and at the national level in Japan, the United States, and West Germany. To cite one example, in 1981 MITI began its Research and Development Project of Basic Technologies for Future Industries. Some of the properties targeted in this project for SiC and Si_3N_4 materials are shown in Table 12.5. Once this project is successfully completed,

Table 12.4. Applications of Engineering Ceramics

Main characteristics	Applications
High-temperature strength, heat resistance	Gas turbine engine parts Diesel engine parts Transport rollers and rails for furnaces High-temperature gas flow regulators Furnace heat-transfer pipes Heat exchangers, etc.
Corrosion, chemical resistance	Pipes for transport under pressure of molten nonferrous metals Immersion rollers Molding nozzles Reaction pipes for chemical reactors Mechanical seals and valves
Abrasion resistance	Mechanical seals, tools, and dies Bearings
Lightness	Control pipes for hydraulic equipment
Other features	Applications which make use of electrical insulating properties: insulated rollers for induction furnace electrode spacers Applications which make use of high elasticity and low expansion: gauges for precision machine tools

Table 12.5. Target Capabilities for Ceramics in Research for the Basic Technologies in the Next Generation of Industries

Type of material	Use conditions

High-strength	In the same environment after processing for 1,000 hours at 1200°C, in air
	  1. Weibull coefficient, $m \geq 20$
	  2. Tensile strength, 30 kg/mm^2
	Creep strength after 1,000 hours at 1200°C, in air, 10 kg/mm^2

Highly anti-corrosive	In the same environment after processing for 1,000 hours at 1300°C in air
	  1. Weibull coefficient, $m \geq 20$
	  2. Tensile strength, $\sigma \geq 20$ kg/mm^2
	  3. Oxidation, ≤ 1 mg/cm^2

High-precision antiabrasive	In the same environment after processing for 1,000 hours at 800°C, in air
	  1. Weibull coefficient, $m \geq 22$
	  2. Tensile strength, 50 kg/mm^2

practical applications of ceramics are expected in a variety of areas: nuclear fission, spacecraft, alternative energy sources, marine development, high-efficiency geo-thermal generators, heat exchangers, chemical reaction furnaces, and high-precision instruments. To achieve these objectives, however, manufacturing technology must, from the outset, develop evaluation techniques and applications technology to suit ceramics. That is essential to ensure the high level of reliability demanded of a material used in mechanical parts.

D. CHARACTERISTICS AND APPLICATIONS OF VERY TOUGH ZrO_2

The rapid increase in the use of ceramic materials for their mechanical capabilities has been brought about by the development of the high-temperature, high-strength materials Si_3N_4 and SiC. With a fracture toughness K_{IC} of only 5–6 MN/m$^{3/2}$, however, they cannot begin to approach the high level of toughness found in metals. The solution to the problem of brittle ceramics is, of course, the achievement of very tough properties in partially stabilized ZrO_2 (PSZ). A technique that utilizes the martensitic transformation of the crystalline form of ZrO_2 from tetragonal (the high-temperature form) to monoclinic (the low-temperature form) (Kobayashi and Masaki, 1982, p. 427) has led to the achievement of high values of toughness, of at least 10 MN/m$^{3/2}$, not hitherto observed in ceramic materials (Kobayashi and Masaki, 1982, p. 427; Mueller, 1982, p. 588). This effect is presented graphically in Figure 12.9.

Some examples of the use to which this very tough ZrO_2 can be put include cutting tools, cutters, knives, ball mill balls, dies, nozzles, and sleeve parts (Kobayashi and Masaki, 1982, p. 427). A problem with PSZ, however, is that the pure stabilized-phase tetragonal ZrO_2 switches to the stabilized single monoclinic form at the relatively low temperatures of a few hundred degrees Celsius, at which time the material loses its high-performance properties. It is to be hoped that more research will be devoted to improving high-temperature properties, so that it will one day be possible to use these very tough ceramics at temperatures above 1000°C.

E. HOW MUCH CAN STRENGTH BE IMPROVED?

The discussion thus far has focused on the development of high-strength, very tough ceramics such as Si_3N_4, SiC, and PSZ. Let us now look, in broad terms, at the potential for improving the strength of ceramics in the future.

The strength of ceramics, a brittle material, is, as is well known, found with the equation $\sigma_C = K_{IC}/Y\sqrt{C}$, where Y is the coefficient for the form of the object and C is the size of the defects (Abe, 1982a, p. 940). To

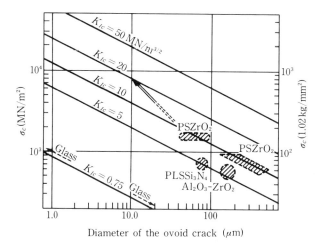

Figure 12.9. Strength of ceramics—present and future. Strength measurements indicated by oblique lines are for flexural strength. $PSZrO_2$ is partially stabilized zirconia; $PLSSi_3N_4$ is pressureless-sintered silicon nitride. From Abe (1982, p. 940).

develop high-strength ceramics, the fracture toughness should be increased and C reduced. The tensile strength σ_C was calculated for ceramics with a variety of fracture toughness K_{IC}, each piece having a round defect of diameter C, to carry out a more quantitative study of the fracture toughness-defect relationship. The results are shown in Figure 12.9 (Abe, 1982, p. 940). The figure presents data for glass as well as ceramics. Since the data did not include every value for σ_C, C, and K_{IC}, the graph was plotted with an admixture of inference.

At present, the effort to develop ceramics with high strength properties has achieved a maximum strength on the order of $\sigma_C \approx 160$ kg/mm^2 (≈ 1600 MN/m^2) $K_{IC} \approx 10$–15 MN/m$^{3/2}$. As can be seen from Figure 12.9, even the strongest ceramics now extent have latent defects of surprisingly large size, with dimensions of several tens of micrometers. It should also be borne in mind that all the strength measurements quoted are measurements of flexural strength. The immediate objective for further development is producing a ceramic material with $C \approx 10$ μm, $K_{IC} = 10$–20 MN/m$^{3/2}$, that is, $\sigma_c = 400$–800 kg/mm^2 (4,000–8,000 MN/m^2), under bending test conditions. When the Weibull coefficient $m = 20$–30 is added, it is obvious that a great deal of effort is required to achieve the stated goals. For glass, a value for σ_c of several hundred kilograms per square millimeter has been achieved (Hara, 1973, p. 784), but because in it K_{IC}

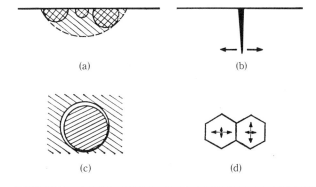

Figure 12.10. Factors that hinder the size C reduction of a defect. (a) shows the multiplied result of small defects; (b) shows the residual stress at the start of a crack (essentially the same as an increase in C); (c) shows cracks that accompany second phase and nonuniformity; (d) shows growth of cracks and residual stress accompanying crystal anisotropy. From Abe (1982, p. 940).

has the low value of $0.75 \text{ MN/m}^{3/2}$, special conditions such as hydrogen fluoride etching are employed to reduce the size of C. In the case of ceramics, however, various obstructive factors are present, so that unanticipated difficulties arise when attempts are made to decrease C (see Figure 12.10).

In pursuing the development of high-strength ceramics, it is essential to clarify the correlation between the performance and properties we are trying to improve, on the one hand, and the character of these ceramics as materials, on the other. Also, the most appropriate technology (in processing the raw materials, forming, firing, and finish processing) must be used to achieve these objectives.

III. CERAMIC PARTS FOR AUTOMOBILES

A. INTRODUCTION

Ceramics have long been used in automobiles in spark plug insulators and other applications, but the percentage of ceramic parts in automobiles has so far been insignificant. Recently, however, the features required of cars have become more diversified, making necessary a wider usage of ceramics. Some examples include oxygen sensors for catalytic convertors to render exhaust nontoxic, honeycomb catalyst carriers, thermosensors for catalyzer protection, knock sensors to improve fuel consumption, riser heaters, and mechanical seals for water pumps. Some of these parts can fulfill other functions apart from the ones anticipated, and thus their significance has increased. For instance, riser heaters effectively improve fuel consumption rates and, at the same time, make the exhaust nontoxic. Similarly, oxygen sensors are effective in improving fuel consumption rates and, at the same time, reducing exhaust toxicity.

The employment of ceramics in automobiles is in its infancy, but this material is already becoming increasingly useful. In the near future, ceramics will find a variety of applications in automobiles.

The operating environment of an automobile is extremely demanding. Ceramics must possess excellent mechanical properties to be used there. The following requirements must be fulfilled:

1. Withstanding 20 g of vibration for extended periods of time
2. For ignition and exhaust parts, withstanding, over an extended period of time, rapid heating and cooling of 50° to 60°C/sec
3. Durable and reliable mechanical characteristics, with a failure probability on the order of 10^{-5} or less

Beyond mechanical capabilities, the following are also necessary:

4. Mass producibility
5. The ability to form reliable joints with metals or with other ceramics
6. Inexpensiveness

Good welding, adhesion, and interference fitting traits are required for joints (item 5 above), as it is normally necessary for ceramic parts to be joined to metal in places. For condition 2, the engine is assumed to be of the reciprocal type, but with a gas turbine engine, the capability to withstand ambient temperature changes of a maximum of 500°C/sec is required. This requirement is approximately equivalent to the condition set in 2, but demands can be greater depending on operating conditions and parts. A high order of ability to withstand creep is required. For instance, with gas turbine generators of high efficiency, as in the "Moonlight" project of Japan's Agency of Industrial Science and Technology, a creep breaking strength of 25 kg/mm^2 or above at 1500°C for 1,000 hours is required (Ueyama, 1979), and similar values are needed for automobiles. Moreover, other capabilities in addition to 1 to 3 are required for some parts.

The parts mentioned above satisfy these requirements and are actually being used in automobiles. Various other engine parts will appear in the future, and are described briefly below.

B. CERAMIC PRODUCTS

1. Oxygen Sensors

An oxygen sensor is used to measure the O_2 concentration in exhaust fumes. The measured concentration is fed back into the engine's air and fuel supply, maintaining complete combustion and making the exhaust gas nontoxic. Zirconium oxide and titanium oxide are used in oxygen sensors. Zirconium oxide (zirconia) is used in tube form; the oxygen concentration in the exhaust is detected through the electromotive force set up between the inside and the outside of the tube. The titanium oxide is used in pellet form, with the oxygen concentration detected through electric conductivity. The structure of the zirconium oxide oxygen sensor is shown in Figure 12.11.

Zirconium oxide for oxygen sensors, as can be seen from the figure, is supported in a way similar to a cantilever. Tensile stress will be applied on the support under 20 g of acceleration; the support must possess the strength to withstand that stress. However, in the structure sketched in Figure 12.11, the stress is extremely small (less than 1 kg/mm^2) and, therefore, ordinarily does not cause a serious problem. In oxygen sensors, the problem is rather one of thermal stress, due to rapid heating and cooling at engine ignition, acceleration, and deceleration. In these cases, the maximum ambient temperature change is 500°C/sec, causing about 40 kg/mm^2 surface tension stress. Therefore, strength which exceeds that level is required of the material. Zirconium oxide is used in the form of stabilized zirconia, which is composed of cubic crystals, or partially stabilized zirconia, which is composed of tetragonal, cubic, and monoclinic crystals. Partially stabilized zirconia has superior mechanical properties at ambient temperatures (see Figure 12.12). Thus, while stabilized zirconia was used in oxygen sensors when initially developed, recently partially stabilized zirconia has been replacing it.

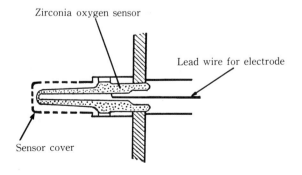

Figure 12.11. Structure of a ZrO_2 oxygen sensor.

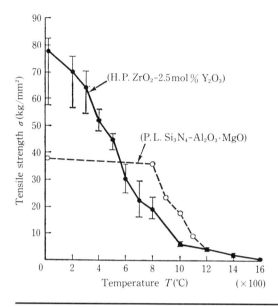

Figure 12.12. Tensile strength of hot-pressed PSZ.

Zirconia oxygen sensors are highly reliable, and, at the very least, almost perfectly satisfy the requirement of reliability for ceramic parts.

2. Honeycomb Catalyst Carriers

Catalyzers are used to detoxify exhaust gases. Pellet carriers and honeycomb carriers (Figure 12.13) are used to bear the catalyzers. Pellet carriers are easy to produce and are unbreakable—and thus highly reliable—because they are made up of small grains. However, since their resistance to air flow and their heat capacity are high, the pellet-type carrier presents some disadvantages when the engine is cold-started. Honeycomb carriers have little resistance to air flow and small heat capacity and offer superior properties in cold start. Thus, the trend has been to switch to the honeycomb type.

Their function dictates that honeycomb catalyst carriers be of thin-walled construction and full of holes. The flexural strength of the material is only about 10 kg/mm^2. These catalyst carriers are operated under the same conditions of heat and vibration as oxygen sensors, and, as in oxygen sensors, the thermal stress is very high. A misfire in the engine may cause localized heating on the surface of the catalyst carriers through afterburning on the surface. Thus, the temperature changes due to rapid heating and cooling on the carriers are at times more severe than on oxygen sensors. Yet the thermal stress occurring in a honeycomb carrier is less than in an

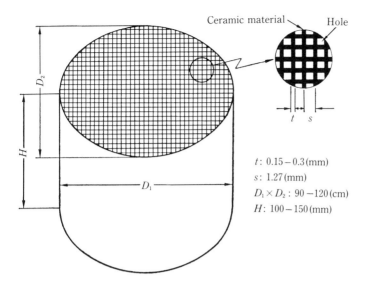

Figure 12.13. Structure and dimensions of a honeycomb catalyst carrier.

oxygen sensor because the entire structure is designed for low thermal expansion, by the use of cordierite ($Mg_2Al_4Si_5O_{18}$) in a wholly thin-walled construction. Thermal stress is roughly evaluated at $\sigma_T \approx E\alpha \Delta T$, where E is Young's modulus, α_T the thermal expansion coefficient, and ΔT the temperature difference. Since E is $2 \times 10^4/mm^2$, and ΔT is 200°–300°C at the front and rear edges along the gas flow through the honeycomb, for σ_T not to exceed the breaking stress, α must be less than 1×10^{-6}°C. To achieve this level of thermal expansion, several approaches have been taken, including devising ingenious composite materials and aligning crystal axes (Lachman, Bagley, and Lewis, 1981, p. 202). These approaches have produced honeycomb carriers which satisfy the demands placed on them as automobile parts.

3. Thermo Sensors

Thermo sensors are provided to prevent deterioration of the performance of catalysts in a catalytic convertor system through exposure to excessively high temperatures. These sensors are mainly made of transitional metallic oxides and placed in suitable metal containers.

The most common shape is the very small ($3–5\phi \times 5–8t$) pellet type. The sensor is completely enclosed in a metal tube and is fixed in inert oxide powder to protect the lead wire, which carries the electric signal from the sensor, from breaking. The most intense stress is placed at the join between the sensor and the metal of the lead. The stresses are caused partly by the shrinkage difference in sintering the sensor between the oxide and the lead metal and partly by unequal thermal expansion between the lead metal and the oxide sensor. These stresses are alleviated by using a metal with sufficient plasticity.

4. Riser Heaters

A heating device or riser heater is placed at the intake to warm the intake air and to carry out evaporative mixing of the fuel completely, to help reach complete combustion when the engine is cold-started. Barium titanate PTC is used for the riser heater for full temperature control in the intake heating, and to make the device more reliable. The riser heater provides nearly perfect combustion as soon as the engine starts, making the exhaust nontoxic and improving fuel consumption ration.

Riser heaters are made from ceramic plates with a metal back up. The metal backing takes most of the stress, and the parts are not exposed to high-temperature exhaust, from the point of view of high-temperature strength. Considerable static stress, however, is caused due to the metal backup construction. Since high stress occurs locally, the ceramic material must have a strength of more than 10 kg/mm^2. Since PTC semiconductivity often occurs at temperatures which facilitate grain growth, its mechanical qualities tend to be lowered. Suitable additives are used to avoid this difficulty. The PTC available on today's market has a mechanical strength (flexural strength) of about 15 kg/mm^2 with crystal grain size of 5–30 μm, and a fracture toughness of 1 MNm$^{-3/2}$. The PTC is required, in addition to mechanical capabilities, to be proof against environmental conditions (able to withstand reduction by fuel, and with stable electrical resistance), and its terminals must be humidity-proof. A balance between mechanical capabilities and these other features must be borne in mind.

5. Knock Sensors

Fuel ignition in the engine should always be adjusted to occur at the upper dead center position of the piston action, to improve the fuel consumption ratio. If the fuel supply parameters are fixed, however, knocking will occur or ignition will occur when the piston is well away from its upper dead center point. The result is possible loss of fuel economy. If knocking is constantly monitored and the vehicle is driven under conditions which immediately precede knocking, however, fuel consumption ratios will be improved. Since knocking is accompanied by a characteristic high-frequency vibration (about 8 kHz) in the cylinder, it is sufficient to monitor that vibration. An oscillating tip is fitted to detect the vibration. The vibrations of the tip are in turn picked up by a resonant or nonresonant oscillator-type sensor. In one method, a piezoelectric ceramic (knock sensor) is empolyed. Its structure is shown in Figure 12.14.

Knock sensors must have extended fatigue life because they are subject to repeated stress, since they use the vibrations of the piezoelectric element. Because one end of the piezoelectric element is fixed to metal, the fixed part is liable to bear high stresses, which it must be strong enough to withstand. Requirements include a suitable value of Q and stability of Young's modulus and the electrodynamic coupling coefficient over a long period of time.

Currently, PZT is used in knock sensors. Stability over several thousand hours has been achieved with fracture toughness of about $6.5 \text{ MNm}^{-3/2}$ and high strength, reaching about 10 kg/mm^2. This material fully performs its function in improving fuel consumption.

C. PARTS IN THE FUTURE

Among automobile ceramics for reciprocal engines, the following developments are expected in the future: precombustion chambers, piston heads, cylinder liners, cylinder heads, and rotors for superchargers. Silicon carbides, nitrides, and partially stabilized zirconia are expected to be used. Flexural strengths of 50 kg/mm^2 or more and thermal expansion coefficients of less than or approximately equal to $3.5 \times 10^{-6} \text{°C}$ will be required as thermal and mechanical properties; the materials mentioned above are close to satisfying these requirements. In particular, partially stabilized zirconia has not only a strength of 100 kg/mm^2, but also far less heat conductivity than other ceramics ($\approx 0.04 \text{ kcal/cm} \cdot \text{sec} \cdot \text{°C}$). It is a promising engine material for heat insulation although it has a tendency to be attacked by hot vapor. Research is particularly pressing ahead into its applications in cylinders and piston heads.

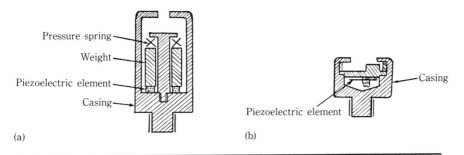

Figure 12.14. Structure of a knock sensor. (a) shows the nonresonant type, and (b) shows the resonant type.

With gas turbine engines to be used in automobiles, the turbine inlet temperature will be about 1350°C to achieve the same ratio of fuel consumption as diesel engines. As of now, only silicon carbide and silicon nitride can withstand such temperatures and are therefore highly promising materials. Properties required for safety are a flexural strength of about 50 kg/mm^2 at 1350°C and about 70 kg/mm^2 at 1000°C for turbine rotors and stators. Creep deformation of not more than 0.5% after 2,000 hours under 12 kg/mm^2 at 1350°C is also required. These requirements are nearly satisfied. In the near future, the urgent task is to establish a manufacturing process which does not harm these qualities of the ceramic material.

Practical applications of ceramics in automobiles are just taking their first steps. No one now has enough experience to know what ceramics should be used where. In the future, as experience accumulates, applications will broaden steadily, establishing ceramics as an industrial material.

IV. ENGINES

For a heat engine, the higher the cycle temperature and cycle pressure, the higher the thermal efficiency rate. Thus, the development of materials which can withstand high temperatures is desirable. At present, Ni- or Co-based alloys, which have the highest levels of strength at high temperatures among high-temperature alloys used, are employed in the hot parts of gas turbines. The turbine inlet temperature of almost 1400°C is achieved by air cooling. Metal materials still have a melting point limitation even with progress in these high-temperature alloys and with improved cooling technologies; recently it has become difficult to hope for any major increase in operating gas temperatures. Therefore, ceramics, with far higher melting points than metal materials can be expected to be used as high-temperature materials in the future. It is predicted and greatly hoped that such materials will have reached the stage of practical use between 1990 and 2000, as shown in Figure 12.15 (Signovelli and Blankenship, 1978, p. 187).

A. CERAMIC GAS TURBINE RESEARCH AND DEVELOPMENT

Applied research projects on gas turbines include Ford's 200 HP gas turbine for automobiles, Cumming's adiabatic engine, the AGT (Advanced Gas Turbine) 100 and 101, as well as CATE (Ceramic Applications in Turbine Engines) projects being carried out principally by the U. S. Department of Energy and NASA, and applied research and development in West Germany and Sweden.

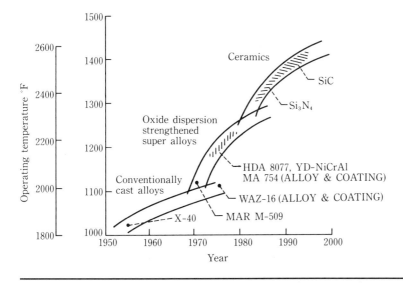

Figure 12.15. Status and development projections for turbine materials. From Signovelli and Blankenship (1978).

The development planning and status of DOE-related projects is shown in Table 12.6 (Masaki, 1983). With these projects, some successful tests of individual parts have been carried out, but systems as a whole have not yet been completed. The technology to replace the high-temperature alloys in current use is the most difficult and high-order problem among the various ceramics applications targeted. Given the present phase of ceramics development, it will be necessary to solve many problems before these goals can be achieved. With respect to the applicability of technology to the gas turbine engine at its present level, methods of application which exploit the special capabilities of ceramics are being devised and put to practical use instead of simply using ceramics as strong materials which have the ceramic properties of thermal, corrosion, and heat resistance. Examples of practical applications are given below.

B. USES IN CORES IN MOLDING AIR-COOLED TURBINE BLADES

In the molding of air-cooled turbine blades, a unidirectional solid blade has recently been made practicable; between 30 and 60 minutes are required for solidification (Ōmi, 1978). For these purposes, the upgrading of ceramics quality is necessary to eliminate inferior products resulting from chemical attack by molding sand, or from deformation and slippage of the

Table 12.6. Ceramic Gas Turbine Development Projects[a,b]

		U.S. projects			West German projects	Swedish projects
		AGT 100	AGT 101	CATE		
Project and target values	Sponsor	DOE	DOE	DOE-NASA	BMFT	United Turbine of Sweden
	Chief contractor	GM DDA	Garrett	GM DDA	VW, DB, MTU	
	Subcontractors	Carborundum, Corning, GTE Lab.	ACC, Ford, Corning, Carborundum	Virtually identical to AGT 100	Degussa, Rosenthal, Starck, ESK	
	Engine type	Two shaft (Gasier/power)	Single shaft	Two shaft (Gasier/power)	Two shaft (Gasier/power)	Three shaft (Gasier/two power)
	Output (HP)	100	100	342	130(VW),200 (DB),400(MTU)	
	Turbine inlet temperature (°C)	1288	1371	1241	1371	
	Turbine type, rpm	Radial flow, integrated, 86,600	Radial flow, 100,000	Axial flow, 37,000? hybrid	Axial flow, hybrid, integrated	Axial flow, integrated
	Fuel consumption (km/1)	18.1	18.2			
	Exhaust temperature (°C)	288	266			

Development status	SαSC(IM) SSN(IM)	SRBSN (SC)	SαSC (IM) SSN (IM)	SN(NP)	SN (IM → HIP)
Rotor, material (fabrication method) Test results	97,100rpm[d]	102,000rpm[d]	56,000rpm[d] 36,000rpm (1,132°C)		
Stator, material (fabrication method) Test results	SαSC(IM) 1,079°C × 40 h	SRBSN 1,150°C × 122 h	RBSC 1,132°C × 894 h	SN(HP),SC(HP)	
Combustor, material (fabrication method) Test results	SαSC(SC,IP)	SαSC(SC,IP)	SαSC	RBSN	
Scroll, material (fabrication method) Test results	1,079°C × 132 h SαSC(SC,IP)	871°C × 107 h SαSC, RBSN	SαSC(IP,IM)		
Heat exchanger, material (fabrication method) Test results	1,079°C × 100 h AS	AS, MAS	1,132°C × 894 h AS	SN	
Seal plate, material (fabrication method) Test results	1,000°C × 2,800 h LAS(SC)	1,093°C × 21 h LAS(SC)	1,000°C × 3,050 h LAS		
Test results	1,079°C × 20 h	1,150°C × 38 h	1,132°C × 1,578 h		

[a] Abbreviations. Materials: SαSC, pressureless-sintered SiC; SRBSN, reaction-bonded resintered Si_3N_4; SSN, pressureless-sintered Si_3N_4; AS, aluminosilicate; LAS, lithium aluminosilicate; MAS, magnesium aluminosilicate. Fabrication method: IM, injection molding; SC, slip casting; IP, rubber pressing, isostatic pressing; HP, hot pressing; HIP, hot isostatic pressing.
[b] From Masaki (1983, p. 53).
[c] Axial flow, integrated.
[d] Room temperatures.

core. Examples of cores for air-cooled turbine blades are shown in Figure 12.16. These cores differ in their characteristics from usual refractory materials, for they are special ceramics which satisfy all the following conditions:

1. Cores can be easily removed chemically from cast parts.
2. Heat expansion of the molded products is equal or similar to that of the mold.
3. Precision of molded parts is high.
4. The mechanical strength to withstand the wax injection pressure or the molten metal casting pressure is necessary.
5. Resistance to heat is high; cores are not reactive with molten metal.

Moreover, following the development of unidirectional solid blades, single-crystal cooled turbine blades are approaching practicability, making the development of higher quality ceramic cores a necessity.

Figure 12.16. Ceramic cores used in air-cooled turbines.

C. HEAT-RESISTANT AND THERMAL BARRIER COATINGS

For some parts of the gas turbine which are exposed to high temperatures, such as combustor lines and a part of the first turbine nozzle blade, thermal barrier coatings have been adopted for heat and corrosion resistance. Both outer and inner surfaces of the cooling blades have ceramic coatings to prolong their life spans. At present, the combustor lines are coated with zirconia heat-resistant material by a plasma spray technique. To minimize the differences in thermal expansion between this coating material and the base material, the percentage of the metal component in the base is increased and coating thickness is limited to between 0.3 and 0.4 mm by a graded coating method which changes the composition step-by-step while approaching the outer surface.

When this technology is applied in cooled turbines, the air cooling rate can be decreased. At the same time, as the thermal flow rate decreases and thermal stress on the blade declines, the life span of the air-cooled turbine is extended. However, temperatures at the outer surface of the ceramic coating increase and absorption of the thermal expansion difference between the metal and the ceramic becomes difficult, so that the part breaks away easily at the joint. This thermal barrier coating technology is a critical and urgent research issue in energy-saving technology. In an example of an application to an actual engine, results of a 300-hour endurance test with a test blade using an adiabatic coating of partially stabilized zirconia containing 8% Y_2O_3 at the surface of the blade have been reported. According to the results, damage and corrosion were obvious and widespread on the blades lacking a thermal barrier coating as in Figure 12.17, but coated blades were not visibly affected. Thus, the thermal barrier coating makes it possible to extend the life of the blade.

D. APPLICATIONS IN CARBON SEALS

In the labyrinth seals widely used for the oil seals of jet engines in the past, a considerable volume of lubricating oil and seal air leaks from the seal (Povinelli, 1975). The great weight and volume of the larger scavenge pumps and piping sizes required thus place restrictions on engine design. Therefore, the superior carbon seals developed recently have been widely adopted, as shown in Figure 12.18. Since this sealing is attached to surfaces and sides at low pressure with an intervening oil film and air film as with a bearing, there is little friction. Moreover, there is very little leakage of lubricating oil compared with the conventional metal labyrinth seals. As can be seen from the figure, carbon seals are set to be attached lightly with springs to respond to the relative movement of the shaft. An oil passage is provided to supply both cooling and lubricating oil.

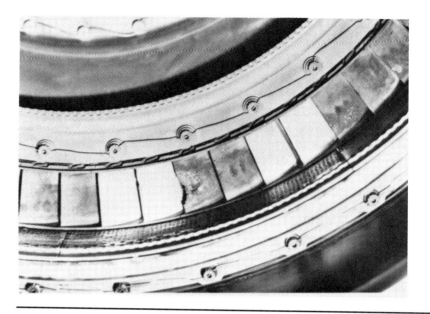

Figure 12.17. Damage to turbine nozzle vanes after 150 hours of endurance testing.

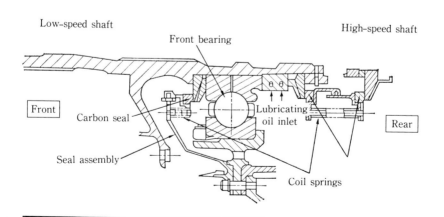

Figure 12.18. Carbon seals in front bearing section of a high bypass ratio jet engine.

E. TURBINE SHROUDS

Decreasing cooling-air consumption and upgrading cycle capabilities by using heat-resistant ceramics instead of metals as in the past is being considered. There are plans to use heat-resistant ceramic shrouds in the alternative energy project for American jumbo jet airplanes (United Technology, 1979).

F. THE STATE OF CERAMICS AS A HIGH-TEMPERATURE MATERIAL

Here let us consider the strength characteristics of ceramics as a high-temperature material, at their present stage of development. Since the strength of an industrial material is generally expressed in terms of tensile strength, ceramics were compared with metallic materials on the basis of results of tensile tests, principally carried out by the authors. Figure 12.19 indicates the variation in the tensile strength of materials with temperature changes from ordinary room temperature to high temperatures. Even ceramics, which are expected to perform as high-temperature and high-strength materials, suffer a significant drop in strength beyond 800–1000°C

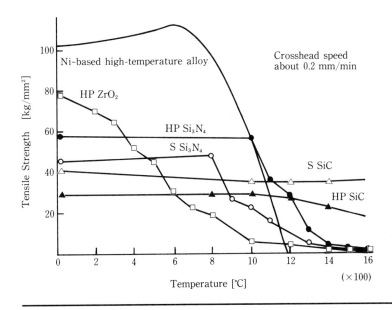

Figure 12.19. Relationship between temperature and tensile strength.

Table 12.7. Creep Strength Results

Nature and dimensions of tested parts	Creep characteristics (rupture time)
Hot-pressed $Si_3N_4{}^a$	1,000 hours at 1400°C in Ar atm; 10 kg/mm^2
3 × 3 × 40 mm	2,000 hours at 1400°C in Ar atm; 5 kg/mm^2
Hot-pressed $Si_3N_4{}^b$	1,000 hours at 1200°C in air; 10 kg/mm^2
3.2 × 3.2 × 13 mm	1,000 hours at 1300°C in air; 5 kg/mm^2
Pressureless-sintered $Si_3N_4{}^c$	1,000 hours at 1200°C in air; 3 kg/mm^2
7ϕ × 36 mm	1,000 hours at 1000°C in air; 10 kg/mm^2
IN100^d	1,000 hours at 850°C in air; 28 kg/mm^2
Ni-based heat-resistant alloy	10,000 hours at 850°C in air; 21 kg/mm^2

[a] From Inoue, et al. (1980, p. 157).
[b] From Govila (1982, p. 15).
[c] From Kawai (1982, p. 206).
[d] From Mechanical Properties Data Center (1968).

because the oxides used as sintering aids are deposited in the grain boundaries and soften at high temperatures. Accordingly, technological development is needed to improve this defect. The results of tests of another high-temperature strength characteristic, creep strength, are shown in Table 12.7. Oxidation at high temperatures is also a problem connected with corrosion resistance. It is a particularly weak point in silicon nitride, which is easily oxidized.

REFERENCES

Abe, Hiroshi. (1982a). *Kyōdo riron to kōkyōdo seramikkusu* [Strength of materials theory and high-strength ceramics]. *Seramikkusu,* 17, 940–950.

Abe, Hiroshi. (1982b). *Seramikkusu no seinōhyōka to shinraisei* [Ceramics: Evaluation of their properties and reliability]. *Kikai gijutsu,* 30, 62–67.

Asahi Glass. (1982). *Enjiniaringu seramikkusu* [Engineering ceramics]. Catalog.

ASTM Standard Book. (1982). Philadelphia: American Society for Testing and Materials, part 10, E399–81.

Hara, Morihisa. (1973). *Garasu no kyōdo no riron* [Theory of the strength of glass]. *Seramikkusu,* 8, 784–789.

Inoue, Hiroshi et al. (1980). Si_3N_4 *Shokettai no kōon hippari kurīpu* [Tensile creep of Si_3N_4 sinterns at high temperatures]. In *Yōgyō kyōkai nenkai, kōen yokōshū* [Abstracts of the annual meeting of the Ceramic Society of Japan]. Yokohama: Yōgyō kyōkai. p. 157.

Iseki, Takayoshi. (1982). *Seramikkusu no kyōdo* [Strength of ceramics]. *Taikabutsu,* 34, 605–613.

Govilla, R. K. (1982). Uniaxial tensile and flexural stress rupture strength of hot-pressed Si_3N_4. *Journal of the American Ceramic Society,* 65, 1, 15.

Kawashima, Chihiro and Suzuki, Hiroshige. (1964). *Hikinzoku zairyō no kōgakuteki seishitsu* [Engineering characteristics of nonmetallic materials]. Tokyo: Chijin shokan.

Kingery, W. D.; Bowen, H. K.; and Ullmann, D. R. (eds). (1976). *Introduction to Ceramics*. New York: John Wiley & Sons.

Kobayashi, Keisuke and Masaki, Takaki. (1982). *Kōkyōdo kōjinsei jirukonia* [Very strong, very tough zirconia]. *Seramikkusu*, 17, 427–433.

Lachman, I. M.; Bagley, R. D.; and Lewis, R. M. (1981). Thermal expansion of extruded cordierite ceramics. *American Ceramic Society Bulletin*, 60, 202–205.

McKenzie, J. K. (1950). The elastic constants of a solid containing spherical holes. *Proceedings of the Physical Society (London)*, B63, 2–11.

Kawai, Minoru et al. (1982). Si_3N_4 *shokettai no kōon hippari shiken* [Tensile tests at high temperatures of Si_3N_4 sinterns]. In *Yōgyō kyōkai nenkai, kōen yokōshū* [Abstracts of the annual meeting of the Ceramic Society of Japan]. Kyoto: Yōgyō kyōkai. p. 206.

Masaki, Hideuki (1983). *DOE keiyakusha kaigi ni miru beiō no seramikku gasu tābin no geniō* [The state of the ceramic gas turbine in the U. S. and Europe as seen in the DOE contractors' meeting]. *Seramikkusu*, 18, 52–56.

Matsusue, Katsutoshi; Fujisawa, Yoshiaki; and Takahara, Kitao. (1983). $ZrO_2 = Y_2O_3$ *kei kaatsu shoketsutai no kōon hippari tsuyosa* [Tensile strength at high temperatures of $ZrO_2 - Y_2O_3$ pressure sintered bodies]. *Yōgyō kyōkai-shi*, 91, 49–51.

Mechanical Properties Data Center, Battelle Columbus Laboratory, U. S. Department of Defense. (1968). *Aerospace Structural Metal Handbooks*. Vol. 5, Code 4212.

Mueller, James I. (1982). Handicapping the world's derby for advanced ceramics. *American Ceramic Society Bulletin*, 61, no. 8, 588–590.

Nakayama, Junn. (1977). *Netsushōgekisei* [Thermal shock properties]. *Seramikkusu*, 12, 144–152.

Ōmi, Toshiaki. (1978). *Kōon gasu tābin buhin no seimitsu kōzō no genjō to shorai* [Present state and future prospects of precision production of parts for high temperature gas turbines]. *Dai 5 kai gasu taabin seminaa shiryōshū*. Tokyo: Nippon gasu taabin gakkai, 4-1-4-10.

Povinelli, V. P., Jr. (1975). Current seal designs and future requirements for turbine seals and bearings. *Journal of Aircraft*, Vol. 12, no. 4, 266–273.

Richerd, D. W. (1982). *Modern Ceramic Engineering*. New York: Marcel Dekker.

Rooke, D. P. and Cartwrite, D. J. (1974). *Stress Intensity Factors*. London: Her Majesty's Stationery Office.

Signovelli, R. A. and Blankenship, C. P. (1978). *NASA CP 2036*. NASA.

Spriggs, R. M. (1961). Heat conductivity process in glass. *Journal of the American Ceramic Society*, 44, 302–304.

Tada, Hiroshi; Paris, Paul C.; and Irwin, George R. (1973). *The Stress Analysis of Cracks Hand Book*. Hellertown, PA: Del Research.

Udagawa, Shigekazu et al. (eds.) (1982). *Fain kemikaruzu toshite no muki keisan-kagōbutsu* [Inorganic silicon oxide compounds as fine chemicals]. Tokyo: Kōdansha.

Ueda, Kanji and Sugita, Tadaaki. (1982). *Seramikkusu no hakai jinsei* [Fracture toughness of ceramics]. *Kinzoku Gakkaihō*, 21, 225–234.

Ueyama, Tatsumi. (1979). *Kōkōritsu gasu tābin* [A high-efficiency gas turbine]. *Kōgyō gijutsu*, Vol. 20, no. 7, 26–30.

United Technology. (1979). *NASA CR-159487*, April.

Wachtman, J. B., Jr., and Lam, D. G., Jr. (1959). Young's modulus of various refractory materials as a function of temperature. *Journal of the American Ceramic Society*, 42, 254–262.

Wakai, Hiroo. (1980). *Jisedai sangyō kiban gijutsu kenkyū kaihatsu* [Research and development on basic technologies for future industries]. *Kogyo rea metaru*, 73, 2–4.

Yamada, Toshiro. (1982). *Kikai sekkei*, 26, 109.

III
MACHINING METHODS

13
Precision Machining Methods for Ceramics

Akira Kobayashi
*Ibaragi Polytechnic College
Mito 310, Japan*

I. INTRODUCTION

Ceramics are superior to metallic or organic materials in many ways, including their great strength and resistance to heat, abrasion, and corrosion. These outstanding characteristics promise a nearly unlimited range of applications.

Advances in electronics technology and space science have led to increasingly stringent requirements for ceramic parts with a high degree of precision. Most ceramic parts, however, are formed from a powder and then sintered at high temperatures. The resulting parts cannot satisfy the requirements for accuracy without precision machining after sintering. Post-sintering precision machining of ceramic parts is now a necessity.

The machining methods presently used in the production of ceramic parts may be classified as removal, modification, addition, joining, and separation. This chapter is concerned with removal methods in precision machining: those methods used to remove unneeded portions from sintered parts, to form parts of particular sizes and shapes, and to obtain pieces with the desired functions.

Table 13.1 presents a classification of removal machining methods according to the type of energy providing the processing. There are many possible methods from which to choose, as the table indicates, and it is necessary to choose the appropriate one for each case in terms of the type of ceramic material being machined and the requirements for surface roughness, form accuracy, productivity, and cost.

A survey of difficulties in the production of ceramics indicates that nearly half the problems raised are related to the machining workability of the material (Nihon kikai kōgyō rengōkai, 1984). The difficulty of improving machining efficiency and accuracy is a major obstacle to the practical use of ceramics.

The results of the Ceramic Society of Japan's survey of problems in removal methods are presented in Table 13.2 (*Ceramics Japan*, 1983). The

Table 13.1. Classification of Removal Methods for Ceramics by Their Energy Source

Source	Method
Mechanical energy	
Cutting	
Single point cutting	Turning
	Shaping
Multiple blade cutting	Drilling
	Milling
Abrasive machining	
Machining with bonded abrasives	Grinding
	Honing
	Superfinishing
	Coated abrasive machining
Machining with loose abrasives	Lapping
	Polishing
	Blasting
	Ultrasonic finishing
	Barrel finishing
	Buffing
Machining by the impact force of water	Liquid honing
Chemical energy	Etching
	Chemical polishing
Electrochemical energy	Electrolytic grinding
Electrophysical energy	Electric discharge machining
	Electron beam machining
	Ion beam machining
	Ion etching
	Plasma machining
Photonic energy	Laser machining
	Laser etching

Table 13.2. Survey of Removal Methods for Ceramics and Their Problems

Method	Machining accuracy	Surface roughness	Nature of the worked surface	Machining efficiency	Mass production techniques	Automation of machining	Performance			Cost of tool wear	Micro-machining	Machining special shapes	Other	All problems
							Machining equipment	Machining tools	Machining costs					
Grinding	72	45	20	106	47	45	49	55	96	73	19	42	2	671
Honing	5	7	3	7	4	2	3	3	3	4	2	2		45
Superfinishing	6	6	3	6	4	3	3	1	4					36
Coated abrasive machining	7	6	2	14	2	4	1	2	6	7	1	2	5	59
Lapping and polishing	32	24	15	40	18	24	8	4	40	14	6	16	2	243
Blasting	6	4		6	4	4	6	2	7	2	4	3		48
Ultrasonic machining	14	2	1	19	12	7	5	8	17	8	9	10		112
Barrel finishing	2	3	2	5	1	3	3	2	5	2			2	30
Buffing	4	11	2	9	4	6	4		5	3		1	1	50
Liquid honing	1	2		2		1	1	2	1	1				11
Chemical machining	2	2	7	4	1	1	2		3	1	4	2		29
Electrolytic machining	1		1	2				1				1		6
Electrolytic polishing	6	4	2	9	5		1		5	2	10	6		50
Electric discharge machining	1													1
Electron beam machining	4	2		3	3		2		4		4	3		25
Ion beam machining	1	1		1		1			1		1			6
Plasma machining	3	2	3	3	5	2			3		3			30
Laser machining	2	1		1	2				1			2		9
Cut-off	65	25	17	88	34	34	34	33	87	81	18	21	3	540
All methods	234	147	78	325	146	137	128	113	288	198	81	109	17	2001

machining methods presenting problems are, in descending order of difficulty, grinding, cut-off, lapping, polishing, ultrasonic machining, and coated abrasive machining. Apart from machining using mechanical energy, interest runs high in electrical discharge machining, laser processing, chemical processing, and ion beam machining. There the most serious difficulties concern productivity, including machining efficiency, suitability for mass production, and automation of the machining process. Next come cost factors, such as machining costs and the cost of the tools consumed in the machining process. The third set of problems concerns the machining accuracy, the surface roughness, and the quality of the machined surface.

Development of the new ceramics has emphasized solely improvement of the characteristics of ceramic materials; almost no thought has been given to improving their machinability. Progress in producing improved characteristics, taking off from the strength and thermal characteristics of ceramics, has been remarkable. But ceramics remain in the category of hard and brittle materials, difficult to machine and presenting many obstacles to precision machining. Thus, successful precision machining must begin with a grasp of the characteristics of each particular type of ceramic material, for only then can the right machining method be chosen and the machining conditions planned appropriately. The following discussion deals with that issue by describing the machining methods presented in Table 13.1, which should be considered with reference to the machining problems summarized in Table 13.2 as well.

II. CUTTING

As problems with materials intractable to machining have increased over the years, a variety of new tool materials have been developed. Figure 13.1 shows one example of the relationship between the hardness of the material to be machined and the hardness of the tool material (Takeyama, 1980). As can be seen from the figure, it is usually necessary to use a tool which is on the order of four times as hard as the material to be machined. That requirement makes diamond an essential material in cutting ceramics.

A. CUTTING PRESINTERED CERAMICS

There have been reports of cutting unfired ceramics with Al_2O_3 (78%), SiO_2 (16%), and CaO (6%) as their principal constituents and ceramics that have been presintered at a range of 500 to 1000°C by using alloy tool steel (SKS 2), sintered carbide (K01), ceramics, cBN, and sintered or natural diamond tools (Narutaki et al., 1984). The differences in cutting characteristics according to the presintering temperature are striking.

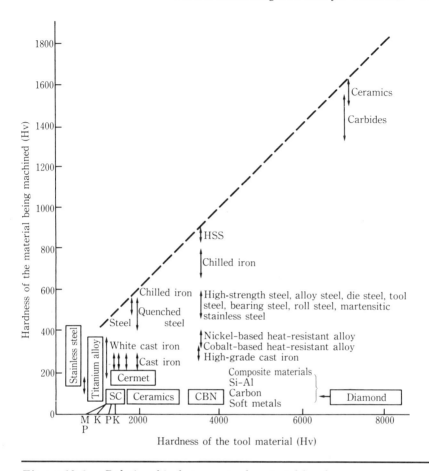

Figure 13.1. Relationship between tool material hardness and hardness of material being machined. From Takeyama (1980).

B. CUTTING SINTERED CERAMICS

The extreme hardness of these substances means that in most cases sintered diamond is used to cut them.

1. Al_2O_3

A tool of sintered diamond (with a grain size of 5 to 10 μm) is used to cut Al_2O_3 of 2,100 to 2,200 Hv. The cutting speed is 30 to 60 m/min, the depth of cut is 1.5 to 2.0 mm, and the feed is 0.05 to 0.12 mm/rev. When cut under wet conditions, a removal rate of 5.0 to 14.4 cm^3/min is obtainable.

A machining efficiency of 3 to 8 times that of grinding with a diamond wheel has been observed for this method (Nakai, 1984). Note that the greater the nose radius of the cutting tool edge, the less wear on the tool. In addition, it has been observed that the main cutting force is small and the normal force is large, which is strikingly different from the relationship between the main cutting force and normal force in cutting metals.

2. Si_3N_4

When a cutting tool of sintered diamond (with a 5 to 10 μm grain size) is used to cut Si_3N_4 of Hv 1400, it has been observed that the dry method gives less tool wear than the wet method and that the lower the cutting speed, the less the tool wear (Nakai, 1984). A machining efficiency of 10 to 32 cm^3/mm was obtained at cutting speeds of 50 to 80 m/min, depth of cut of 1.0 to 2.0 mm, and feed of 0.05 to 0.20 mm/rev.

3. ZrO_2

Zirconia is not very hard and causes little wear on a sintered diamond cutting tool. Milling can also be used to cut this material (Nakai, 1984).

4. SiC

Observations have shown that in cutting SiC of Hv 2000, the dry method produces less wear than the wet method and occasions less cutting force as well (Nakai, 1984). The tool wear, however, is considerably higher than that found when the other ceramics mentioned above are cut.

5. Mica Ceramics

Mica ceramics formed by sintering or fusing fluorine mica are easy to cut (Murai, 1980). There are five types of mica ceramics: mica-layer-sintered, phosphate-bonded, glass-bonded, binderless mica ceramics, and a composite type. Macor, which is marketed by Corning Glass Works, belongs to the binderless type and is very easy to cut (Iwata, 1980). Published results of the drilling of glass-bonded mica ceramics indicate that their machinability worsens as their glass content increases. Anisotropy due to the direction of force during sintering was also reported, as was that machining at a faster rate is associated with better machinability. Futhermore, diamond tools were reported to produce pieces with better accuracy than do high-speed steel or sintered carbide tools (Murai, 1980).

C. ULTRAPRECISION DIAMOND CUTTING TOOLS

Several types of ultraprecision machine tools using highly accurate pneumatic bearing spindles have been developed recently (Kobayashi 1979a,

13. Precision Machining Methods for Ceramics 267

1982a, 1983a, 1985a). With diamond tools attached to these machines, a mirror finish has become obtainable without lapping or polishing. This development is a great contribution to lowering machining costs.

To reach high form accuracy and surface roughness by diamond cutting, at least five conditions must be considered. They are as follows:

1. The diamond cutting tool edge must be as sharp as possible.
2. The movement of the machine tool must be precise.
3. There must be no influence of vibration.
4. There must be complete dynamic balance.
5. Care must be given to temperature control.

These are the basic considerations. Let us take up each one in detail.

1. Diamond cutting tools can provide a sharper edge than can tools of high-speed steel, sintered carbide, ceramics, or other materials. Also, the friction component between tool and work material is small, and it is possible to give a minute depth of cut. Thus, it is possible to transfer and reproduce the machine tool accuracy in the workpiece. In addition, almost no tool wear as built-up edge occur. Thus, such diamond cutting tools have made it possible to achieve ultraprecise cutting accuracy.

The materials which can be cut without causing a built-up edge are limited, however. At present diamond cutting is possible for gold, silver, copper and copper alloy, aluminum and aluminum alloy, electroless nickel, lead, and other soft metals; germanium, silicon, potassium phosphate, sodium chloride, potassium chloride, and other salts; and several types of plastics. It cannot be used to machine steel, titanium, or molybdenum. Recently research on the possibility of mirror surface cutting of glass has been published and is arousing interest (Brehm, van Dun, Teunissen, and Haisma, 1979). When forming diamond tools, the selection of crystalline orientation and the method of polishing the edge are important points.

2. To reproduce the machine tool accuracy in the workpiece, accurate movement of the tool is critical. A great effort has been made for many years to improve the accuracy of the spindles which are the basis of rotary motion and of the slide mechanisms which are the basis of linear motion (Brehm, van Dun, Teunissen, and Haisma, 1979; Whitten and Lewis, 1966; Bryan, Clouser, and Holland, 1967; Warmbrod, 1972; Williams and Warmbrod, 1973; Saito, 1978; Kobayashi, 1978; Bryan, 1979; Suzuki, 1980; Tajima et al., 1981; Bryan, 1981; Sumiya et al., 1981; Usuki et al., 1982; Niwa et al., 1982; Wada, 1982; Tanaka, 1982; Pardue, 1982; Carter, 1982; Donaldson, 1982; Thompson et al., 1982; McKeown, 1982).

3. Vibration originating in the machine tool itself or vibration transmitted from outside will worsen the machine surface roughness obtainable.

Thus, thoroughly investigating the causes of vibration and eliminating them are important.

4. To achieve full dynamic balance, the spindle itself and the face plate to which the workpiece is attached must be in dynamic balance individually; moreover, every effort must be made to ensure the dynamic balance of the combined setup. When a diamond tool is attached to the spindle for cutting in the fly cutting method, it is necessary to exercise the same care.

5. The oil shower system has been developed to ensure accurate temperature control for the machine tool (Warmbrod, 1972). On/off control of solenoid valves maintains the oil temperature at 20 ± 0.006°C while it is showered, inside a plastic cover, at a pressure of 0.28 MPa and a flow rate of 150 L/min over the workpiece being machined and the machine tool. Using this method has contributed greatly to improving the form accuracy of diamond machining. Its use has, therefore, become widespread. The DTM #3 at Lawrence Livermore National Laboratory, University of California, has the shower of oil controlled to ±0.0025°C at a rate of 1.5 m^3/min (Wada and Bull, 1982).

Cutting with diamond tools has many advantages in terms of the surface roughness, form accuracy, and low costs realizable. First, the smoothness of the machined surface is excellent. Using the equipment shown in Figure 13.2 and the fly cutting method to cut oxygen-free copper or plastic, a cutting surface roughness of 4 nm R_{max} is obtainable (Kobayashi, 1978). When equipment with a spindle with excellent rotational accuracy and a sharp-edged diamond cutting tool is used, surface roughness of about 0.01 $\mu m R_{max}$ is relatively easy to achieve. The materials tabulated on p. 269 show surface roughness when ceramics are cut by diamond tools. Among them there are many examples of application to the production of infrared reflectors using polycrystalline germanium.

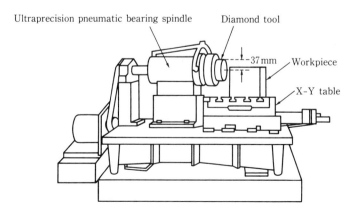

Figure 13.2. *Experimental device for ultra-precision diamond cutting. From Kobayashi (1978).*

SrF$_2$	ZnS	CaF$_2$	NaCl	KCl (RbCl)	KCl (EuCl$_2$)	ZnSe	Ge
4.6	8.1	8.6	9.1	10.2	11.7	15.3	5.6 ~ 27.4

Cutting with diamond tools also permits excellent form accuracy in machining. Apart from the question of machined surface roughness, it is quite difficult to machine large-diameter aspheric surface reflectors with high precision. In making reflectors of soft metals, it is difficult to improve the surface roughness and dimensional accuracy simultaneously by polishing. In this case cutting with diamond tools using high-precision, high-rigidity machine tools has been recognized as clearly superior.

In addition to these advantages, using diamond tools can cut costs. In the machining of reflectors used for observational systems in the infrared region and of refracting parts, the use of diamond cutting has contributed greatly to notable reductions in machining costs.

Since they have so many advantages, a variety of ultraprecise diamond cutting machine tools have been produced, as shown in Table 13.3. Also, cutting with diamond tools has come to be the method of manufacture for reflectors for laser fusion devices. In addition, it is used in the production of optical parts for forward looking infrared radar systems in the infrared region and astronomic telescopes in the ultraviolet and X-ray regions. A rapid expansion of the use of diamond cutting tools for ceramics can be expected.

III. GRINDING

In grinding, the removal process is carried out as an accumulation of minute cuts by the fine cutting edges of the hard abrasives in the surface layer of the grinding wheel. Therefore, in selecting grinding conditions, greater weight should be given to the cutting depth of each individual abrasive's cutting edge (grain depth of cut) rather than to the cutting depths of the grinding wheel. Grinding ceramics, hard and brittle materials, requires particular care in selecting the grain depth of cut.

A. GRINDING MECHANISM

Figure 13.3 shows a typical grinding process. It indicates the momentary interval in cylindrical plunge grinding between when one particulate cutting edge X has separated from the workpiece and the next cutting edge Y is about to separate from it. The hatched area is the cross-sectional form of the chips produced by Y. Diameter (mm) and peripheral speed (m/sec) for the workpiece and the grinding wheel are, respectively, d_ω, d_s and v_ω, v_s. If the depth of cut is a (mm) and the grinding width is b (mm), then PBC,

Table 13.3. History of the Development of Ultraprecision Diamond-Turning Machines[a]

	1960	1965	1970	1975	1980
U.S.A					
Research organizations					
Department of Energy-related					
Union Carbide, Y:12	●──Microinch Machining──→ ○ du Pont No. 2	○ Moore	○ du Pont No. 3 (R.θ Lathe)	○ Multi-facetted Mirror	○ Ex-Cell-O ○ POMA'
	du Pont No. 1	──Spindle accuracy──→			
Lawrence Livermore (National) Laboratory			○ Moor No. 1	○ Moore No. 2	○ Moore No. 3 84* ○ BODTM ● 60'LODTM
			(Oil shower)──→	Oil shower (Laser Interferometer)	
Battelle Pacific Northwest Laboratory				○ Omega-X Nanometer	
Naval Weapons Center				○	
Michelson Laboratory					
Rockwell International				● αθ machine	
Machine tool manufacturers					
Moore Special Tool			△ Union Carbide LLL	△ Bell & Howell LLL	△ M18-AG LLL
Pneumo Precision				──Micro Surface Generator──→ ● MSG-500	△ MSG-325 ○ MSG-700 ○ Polytech 1000
Ex-Cell-O				△ II-C ▲ III-B	
Intop				LASL Union Carbide	●

Firms machining parts
　Perkin-Elmer
　Honeywell
　Bell & Howell
　Polaroid
　Optical Science Group
　Applied Optics Center
Europe
　Research organizations
　Culham Laboratory
　Cranfield Unit for Precision
　　Engineering.
　Rank Taylor Hobson
　Philips Research Laboratory.
Japan
　Machine tool manufacturers
　Toyoda Machine Works
　Hitachi Seiko
　Fujikoshi
　Toshiba Kikai
　Firms machining parts
　Sharp
　Cannon
　Copal Electra

[a] ○, Horizontal spindle type: ●, vertical spindle type: △, production and delivery.

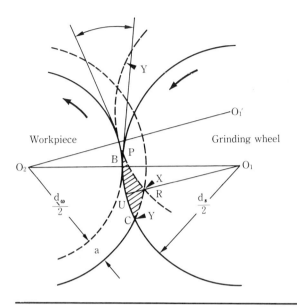

Figure 13.3. A grinding model.

the arc of contact, is given by

$$l_c = \left(\frac{d_s d_\omega}{d_s + d_\omega}\right)^{0.5} a^{0.5} \tag{13.1}$$

RU, the maximum grain depth of cut, is found by

$$h_{max} = 2L_s \left(\frac{v_\omega}{v_s}\right) \left(\frac{d_s + d_\omega}{d_s d_\omega}\right)^{0.5} a$$

$$= 2L_s \left(\frac{v_\omega}{v_s}\right) \frac{a}{l_c} \tag{13.2}$$

where L_s is the distance (mm) between neighboring grains X and Y. The average grain depth of cut is

$$\bar{h} = \frac{h_{max}}{2} = L_s \left(\frac{v_\omega}{v_s}\right) \frac{a}{l_c} = \frac{v_\omega}{v_s} aA \tag{13.3}$$

where $A = L_s/l_c = (Cl_c \bar{b}')^{-1}$
C = the number of effective cutting edges per unit area
$\bar{b}'$ = the average chip thickness

For internal grinding, d_ω is negative. For surface grinding, $d_s = \infty$.

To determine the grinding volume per unit time Z, calculate

$$Z = abv_\omega \tag{13.4}$$

where Z is given in units of mm^3/sec. The grinding volume per unit width and unit time is calculated by

$$Z' = av_\omega \tag{13.5}$$

where Z' is given in units of mm^3/sec·mm.

If V_ω (mm^3) is the stock removal and V_s (mm^3) is the volume of grinding wheel wear, then the grinding ratio

$$G = V_\omega/V_s \tag{13.6}$$

The two components of the grinding force $F(N)$ are represented by the tangential force $F_t(N)$ and the normal force $F_n(N)$. Compute the power required for grinding as

$$P = F_t v_s (W) \tag{13.7}$$

and the specific grinding energy as

$$e = P/Z \tag{13.8}$$

where e is given in units of J/mm^3. It is advisable to use these items as the criteria for selecting the grinding wheel and grinding conditions in grinding ceramics.

B. DIAMOND WHEELS SUITABLE FOR GRINDING CERAMICS

Recently, the use of diamond wheels in grinding ceramics has become widespread (Kobayashi et al, 1954; Kobayashi, 1970; Juchem et al., 1979). With a diamond grinding wheel, the hardness of the abrasive grains is much greater than that of Al_2O_3 or SiC grinding wheels. The smaller percentage of grains (the ratio of abrasive grains contained in the grinding wheel) means that the distance between grains L_s is large and that the pores, which acts to let the chips escape, are few. It is necessary to consider these characteristics carefully.

In the diamond grinding of ceramics, the following conditions are greatly to be desired: (1) a large Z (Eq. (13.4); (2) a small P (Eq. (13.7)), (3) a large G (Eq. (13.6)), and (4) high accuracy in size and shape of the product with low surface roughness and surface damage.

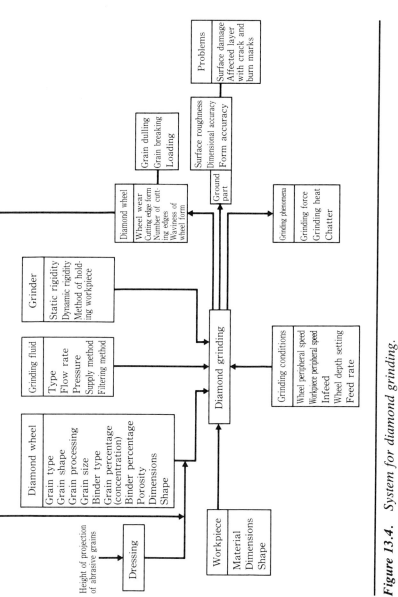

Figure 13.4. System for diamond grinding.

13. Precision Machining Methods for Ceramics   275

To achieve these conditions, it is necessary to give careful thought to the type of grinder, the method of supporting the workpiece, and the rigidity, in terms of the characteristics of the ceramic material, which are different from those of metals (Ishida, 1980). In any case, as Figure 13.4 indicates (Kobayashi, 1970), it is necessary to consider the grinding process as a total system, including the workpiece, the grinding wheel, the grinding fluid, and the grinder. As stated, the grain depth of cut (Eqs. (13.2) and (13.3)) is an important guide in the selection of grinding conditions.

The development of diamond wheels suitable for use with ceramics and of dressing and truing methods for using them is an urgent problem. A number of recommendations have been made for diamond wheels suitable for use with ceramics (Ishida, 1980; Tsujigō, 1983 and 1982; Nakao, et al., 1985), one of which is presented in Table 13.4 (Ishida, 1980). Synthetic diamond grains (identified by the SD mark) are used extensively. The numbers after SD in the table indicate the grain size. The numbers after the hyphen give the concentration. A bronze metal bond (indicated by M) of the Cu–Sn group is widely used as a binder. A resin bond (B) wheel is preferred to avoid chipping while grinding ceramics.

Recently several new types of diamond wheels have been developed to the stage of practical use in grinding ceramics. That presented in Figure 13.5 has a larger percentage of grain than in conventional diamond wheels, and coated diamond particles and special fillers and bonds are used in it (Komine, 1984). When this type of diamond wheel is used for surface grinding of hot-pressed Si_3N_4, it is possible to grind with high efficiency (>15 $mm^3/sec \cdot mm$. The grinding ratio is also said to be greatly improved.

A diamond wheel with porous metal bond has been developed (Tsujigō, 1982); when quartz glass is ground with a straight cup wheel, the dressing intervals are improved by 30%. It has also been observed that there is little chipping at the faces of the workpiece.

The development of a diamond wheel with cast iron bond was recently announced (Nakagawa, et al., 1984). In grinding Si_3N_4, a wheel depth of

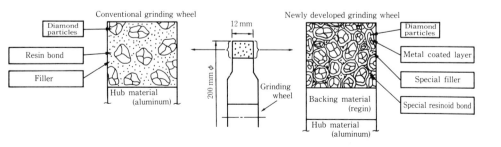

Figure 13.5. Structure of a newly developed diamond wheel for grinding ceramics.

Table 13.4. Recommended Standards for Diamond Wheels for Grinding Ceramics[a]

Workpiece	Operation	Wheel Shape	Wheel Type
Glass	Rough grinding	6A2S, 11A2S	SD120–75M
Optical glass		1A1, 14A1	SD120–75M
Crystal glass	Fine grinding	6A2S, 11A2S	SD270–75M
		1A1, 14A1	SD270–75M
	Fine grinding (cut surface)	14EE1	SD325–100M
	Fine grinding (centering and edging)	1DD1	SD270/325–100M
	Fine grinding (centering and edging)	1DD1	SD325P
	Cut-off	1A1R	SD100–50M
	Fine Cut-off	1A10R	SD270P
	Hole drilling	2A2RS	SD120P
	Edging	1FF6Y	SD140–75M
	Beading	1EE1V	SD500–50M
Ceramics	Rough grinding	6A2S, 11A2S	SD120–75M
		1A1, 14A1	SD120–75M
Alumina refractories	Fine grinding	6A2S, 11A2S	SD270–75M
		1A1, 14A1	SD270–75M
	Cut-off	1A1R	SD100–50M
Quartz	Rough grinding	6A2S, 11A2S	SD120–75M
Artificial quartz	Fine grinding	6A2S, 12A2S	SD270–75M
		1A1, 14A1	SD270–75M
	Cut-off	1A1R	SD100–50M
Silicon	Fine cut-off	1A10R	D325P
		6A2, 11A2	SD500–50BM
	Edging	1EE1V	SD500–50M
Ferrite	Rough grinding	6A2S, 11A2S	SD120–75M
Carbon		1A1, 14A1	
Graphite	Fine grinding	6A2S, 11A2S	SD270–75M
		1A1, 14A1	SD270–75M
	Cut-off	1A1R	SD100–50M
	Fine cut-off	1A10R	SD270P
Cermet	Normal grinding	1A1	SDC200–100B
TiC group		6A2	SDC200–100B
TiN group		11V9	SDC200–100B

[a] From Ishida (1980).

cut of up to about 20 μm was obtained, together with an improved grinding ratio (Ogyūda, et al., 1984). The new type of diamond wheel can also be used in honing and lapping.

When Si_3N_4, SiC, or Al_2O_3 is ground by a diamond wheel of 400 grain size or by a cBN wheel, the ground surface roughness and the grinding ratio are as given in Figure 13.6 (Itō, 1981 and 1983). It is clear that, under the same grinding conditions, there is a great difference in the roughness obtained, depending on the type of work material. In addition, it can be seen that the ground surface roughness realized by grinding with a cBN

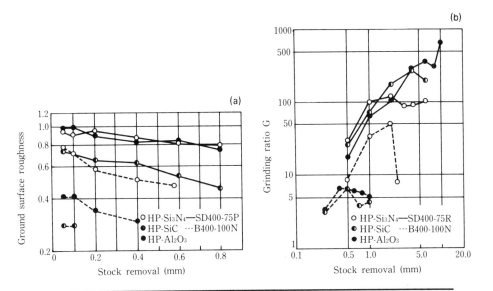

Figure 13.6. Comparison of the ground surface roughness (a) and grinding ratio (b) when several types of ceramics are ground with a diamond and a cBN grinding wheel. From Ito (1981, 1983).

wheel (dashed line) is superior to that realized with a diamond wheel (solid line). It is thought that because the cBN, which is softer than diamond, is worn away relatively quickly and grinds in a glazing state, the ground surface roughness is improved. Studying the grinding ratio data gives one knowledge of one aspect of selection of a grinding wheel suitable for the particular type of work material.

C. MACHINE TOOLS FOR GRINDING CERAMICS

Results have been reported (Miyashita and Yoshioka, 1980) for grinding hard, brittle materials such as quartz, using a surface grinder fitted with an ultrafine cutting device using a high-precision, highly rigid pneumatic bearing spindle, as illustrated in Figure 13.7. A transparent surface was achieved by grinding using a GC 120H grinding wheel and a 0.2 μm step feed. In addition, scratch streaks, judged to be plastic flow by the cutting edges of the abrasive grains, were discerned. Furthermore, use of an SD 1500 N metal-bond diamond wheel to grind the quartz resulted in a surface roughness of 0.1 μm R_{max} and recognizable transparency.

When ultrasmall cuts of 0.1 μm steps are applied in this way, the grinding of hard, brittle materials occurs not by an assemblage of fine cracks but by plastic flow, and a transparent, outstandingly smooth surface may be

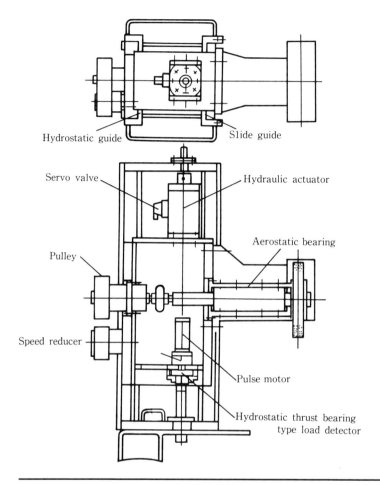

Figure 13.7. Structure of a surface grinder for hard, brittle materials. From Miyashita and Yoshioka (1980).

produced. This conclusion has led to great expectations for the future of grinding hard, brittle materials. It is necessary, therefore, to pour energy into the development of a high-precision, highly rigid grinder suitable for use in the diamond grinding of ceramics. It is essential to consider its structure from a different perspective from that taken for grinding metals, in terms of work spindle, wheel spindle, workpiece support, wheel support, guide surface equipment for controlling wheel depth of cut, and truing equipment (Moritomo et al., 1983; Itō, 1984).

A precision rotary surface grinder for grinding silicon, ferrite, and similar materials has been developed (Usuki et al., 1985). In its development, it was decided that the rough grinding to create the accurate form of the workpiece and the finishing grinding to ensure surface smoothness and a worked layer would be carried out with separate grinding wheels. The rough grinding creates brittle fractures, and it is important to select grinding conditions to make those cracks as shallow as possible. Concretely, the technique is as follows:

1. The machine is designed to have two wheel spindles for use in rough grinding and finishing, for machining with a single chucking.
2. Hydrostatic bearings are used in the wheel spindle and the table slide.
3. The grinding is done using the in-feed method.

An example of results obtained using such a grinder is presented in Table 13.5.

The results of cylindrical plunge grinding of Al_2O_3, ZrO_2, SiC, and Si_3N_4 with a metal bond diamond wheel (Nakajima, 1985) and centerless grinding of Al_2O_3 or B_4C with a resinoid diamond wheel (Daitō, 1985) have also been published.

IV. LAPPING AND POLISHING

Lapping is a machining method which has been used from antiquity to form relatively simple geometrical forms—flat, spherical, or cylindrical surfaces. The method permits production of quite high-precision products even using a machine with relatively low precision. Normally, cast iron or other comparatively hard laps and abrasive grains of several micrometers or more in size are used.

A. THE LAPPING AND POLISHING MECHANISM

The lapping of ceramics is explained by the formation of minute cracks on the work surface, their growth, and the falling off of chips developing on them, as shown in Figure 13.8 (Imanaka, 1967). That is, the abrasive grains which are carried between the lap and the workpiece and share the load cause deformation of the workpiece, in response to the point shape and allotted load. When the set limit is passed, cracking occurs due to tensile stress; as the cracks grow, the material chips off the workpiece. The main factors influencing the lapping characteristics are the type of lap, the type and size of the abrasive grains, the lapping fluid type, the lapping pressure, and the lapping speed. The critical point is to select appropriate values for the type of ceramic to be lapped.

Table 13.5. Machining Examples with a Precision Vertical Surface Grinder[a]

	Workpiece			
	Mn-Zn ferrite single-crystal (110) surface 12 × 13 × 1.5^t		Al_2O_3 50$^{\phi}$ × 33$^{\phi}$ × 10^t	ZrO_2 40$^{\phi}$ × 20$^{\phi}$ × 12^t
Operating conditions	Rough grinding	Finish grinding		
Grinding wheel	SD400P75B7-2	SD1200P200B	SDC200P50B	SD400P75B
Grinding wheel, peripheral speed (m/min)	1200	1200	1200	1200
Table speed (m/min)	1200	1200	1200	1200
Infeed speed (μm/min)	20	1	3	2
Spark out (min)	1	2		
Machined volume (μm)	30	15	10	10
Grinding fluid type	Water-soluble	Water-soluble	Water-soluble	Water-soluble
Surface roughness R_{max} (μm)		0.04	2.5	0.2

[a] From Usiki (1985).

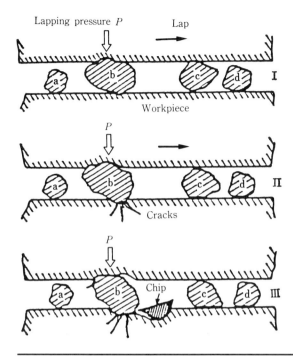

Figure 13.8. Model of the lapping process for hard, brittle materials. From Imanaka (1967).

For instance, Figure 13.9 presents the specific lap volume per unit pressure, the surface roughness, and the depth of damage layer observed in lapping single-crystal $LiTaO_3$ with a range of types of abrasive grains and grain sizes (Kasai, 1978a). Here the ratio number η_w when the lapping volume is proportional to the lapping speed, lapping pressure, and lapping time is called the specific lap volume per unit pressure.

Lapping of high-strength ceramics has a tendency to lower machining efficiency in any case. Also, accuracy in lapping is ensured by transfer of the planar or other geometrical form of the lap to the workpiece. Thus, lapping of high-precision pieces requires a lap maintained at a high level of accuracy in shape. In lapping ceramics, the lap easily abrades away, and thus a variety of methods are used to correct its accuracy.

Since, in lapping ceramics, the machining proceeds due to brittle fractures on the surface, a layer of cracks tends to remain beneath the worked surface. For instance, when polycrystalline ferrite is lapped, the growth of cracks to a depth of about one-fifth of the grain size of the abrasive grains used has been observed (Watanabe et al., 1972).

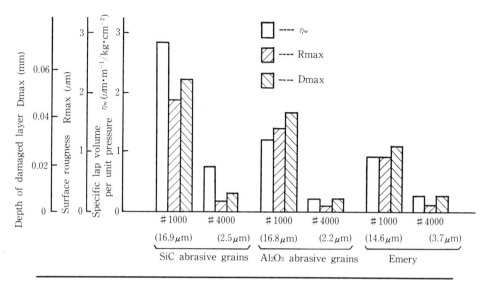

Figure 13.9. Types of abrasive grains and machining characteristics in lapping single-crystal $LiTaO_3$. From Kasai (1978a).

The polishing process is performed after lapping and is intended to smooth the surface while maintaining the precision of form obtained by lapping. Both soft polishers and fine abrasive grains are used, and the polishing proceeds at relatively low pressures. The differences between lapping and polishing lie in the size of the abrasive grains and their holding method (Yasunaga, 1980; Watanabe, 1983). In polishing, the processing occurs due to the fine scratch effect carried out with extremely fine cuts into the workpiece by the grains held elastically and plastically in the soft polisher. There are none of the tiny cracks that occur in lapping. Asphalt, tar pitch, and felt have long been used as polishers, but recently polyurethane and porous nonwoven fabric sheets have also been used. Abrasive grains of diamond, Al_2O_3, Cr_2O_3, CeO_2, and SiO_2 are used.

Polishing creates extremely small irregularities in the worked surface and a slight worked layer. Thus, by polishing it is possible to obtain a higher degree of accuracy in dimensions and shape and greater surface smoothness than with lapping. The machining efficiency, however, is lower.

B. NEW POLISHING METHODS

Recently a host of new polishing methods have been developed, based on a variety of new approaches. These new methods are nearing readiness for practical application. Several of them are discussed below.

1. Complex Polishing

Since it is difficult to ensure form accuracy of the workpiece with chemical polishing, in this method the form accuracy is provided by mechanical polishing with abrasive grains. In addition, machining defects caused by the mechanical removal process using abrasive grains are eliminated by chemical polishing.

A concrete example concerns GGG ($Gd_3Ga_5O_{12}$), for use as magnetic bubble memory elements (Karaki, et al., 1978). A machine with the same structure as a lapping device is used. Instead of a cast iron plate, a disk of bronze to which artificial leather is glued is used, and alkaloid colloidal silica is used as the abrasive grain. The same method may be used for other materials, such as garnet ($Nd_3Ga_5O_{12}$, $Sm_3Ga_5O_{12}$) for substrates. In addition, a method using a liquid with SiO_2 find particles of about 0.01 μm diameter suspended in an alkaloid liquid is widely used in polishing single-crystal silicon and $LiNbO_3$. In polishing optical elements, the elasticity of artificial leather makes sags occur readily and prevents developing complete flatness.

2. Mechanochemical Polishing with Soft Powder

A powder that is softer than the work material, yet is rich in chemical reaction with it, is used as abrasive powder. Under the high temperature and pressure at the point of contact in the machining, a solid-phase reaction occurs. The microreaction portion is removed in microscopic amounts by the resulting frictional force created on it. Thus a mirror surface with few surrounding sags or machining strains can be obtained (Yasunaga, 1975, 1980).

When α-Al_2O_3 single crystals (sapphire) are polished with SiO_2, Fe_3O_4, Fe_2O_3, TiO_2, or other powders with a high reactivity with sapphire, a clearly greater amount of machining is required than with the much harder α-Al_2O_3 or $FeAl_2O_4$, as is also true for the dry method as compared to the wet method. This mechanochemical polishing method makes possible faster polishing than with mechanical polishing using diamond powder.

Similarly, $BaCO_3$, $CaCO_3$, or Fe_3O_4 can be used in polishing single-crystal silicon. In particular, $BaCO_3$ gives a much better machining efficiency than does polishing with SiO_2 or CeO_2, the conventional materials; Fe_3O_4, MgO, and MnO_2 are appropriate for polishing quartz. Particularly excellent machining efficiency is realized by dry-method polishing of quartz using Fe_3O_4 and a copper polisher, and a scratch-free smooth surface is produced (Kasai et al., 1978b). It has been reported that, in mechanochemical polishing of hot-pressed Si_3N_4 with Fe_2O_3 or Fe_3O_4, a machining volume of 1.6 μm/H was achieved with better than 20-nm surface roughness and a worked layer of less than 10 nm (Vora, Orent, and Stokes, 1982).

3. The Elastic Emission Machining Method

This method is an attempt to make possible atomic-order fracture phenomena by mechanical means. The elastic emission machining method uses ultrafine particles of Al_2O_3 and ZrO_2, making them collide with the workpiece surface in the way illustrated in Figure 13.10 (Mori, 1980). This method requires the use of an abrasive powder with ultrafine grain size and uniform grain-size distribution. In addition, to alleviate the force with which the particles press against the workpiece, the tool must be an elastic substance. Finally, it is important to prevent localized, concentrated loads on the workpiece. With the elastic emission machining method, surface roughnesses of about 5 Å for single-crystal silicon have been reported.

4. Noncontact Polishing

As shown in the outline of its principles given in Figure 13.11, in noncontact polishing the polisher surface and workpiece are not allowed to come into direct contact. When a laminar flow of fluid is caused in the gap of 0.5 to 2 μm, the particles are able to remove microscopic amounts from the workpiece. This method entails no wear on the polisher and can permit stable polishing. For 75-mm diameter silicon, a flatness of around 0.3 μm has been achieved (Watanabe et al., 1981).

5. Hydration Machining

This method aims at producing a mirror finish by removal machining without the use of abrasive grains, through the mechanical action of the polisher on the microscopic products of reaction resulting from hydration that occurs in a superheated steam atmosphere (Okutomi et al., 1979).

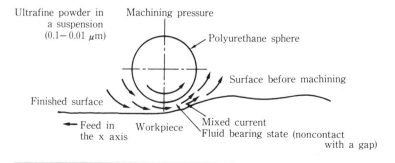

Figure 13.10. Diagram of the elastic emission machining (EEM) principle. From Mori (1980).

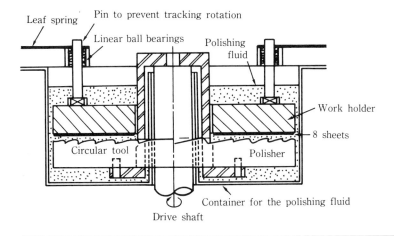

Figure 13.11. Structure of a noncontact polishing device. From Watanabe et al. (1981).

Precision machining of Al_2O_3-group compounds, MgO, glass, and ZnSe crystals is possible with mirror finish roughness on the order of 0.01 μm.

6. Noncontamination Machining

To reduce contamination by elements other than those of the ceramic substance being machined, the aim of this method is to achieve a mirror finish by abrasion with a ceramic of the same composition as the workpiece ceramic (Sugita, 1978). High form accuracy and outstanding workability are achieved with this method.

V. LASER PROCESSING

A. DISTINCTIVE FEATURES

The most important feature of laser light is its spatially coherent beam. It has excellent monochromaticity and phase alignment. The next feature is its directionality; the angle of divergence is unusually small. Thus, when a laser is focused using optical systems, it is possible to squeeze down to a quite small spot diameter. Also, when the power density on the surface of that spot is calculated, it is found to vary over a considerable range but can reach values of power density equaling the highest obtainable by conventional processing methods.

Table 13.6. Types of Lasers Used in Material Processing

Type	Wavelength (μm)	Operating mode	Typical power (W)	Main applications
Solid state				
Ruby	0.6943	Pulsed	10^5 (peak)	Spot welding, drilling
YAG	1.063	(Continuous) CW	300 or less	Welding
		Q-switched	5×10^3 (peak)	Trimming
		(Repeating) pulsed	200 or less (average)	Welding, drilling
Glass	1.063	Pulsed	10^6 (peak)	Spot welding, drilling
Gas				
CO_2	10.6	(Continuous) CW	15×10^3 or less	Heat treating, welding, cut-off
		Repeating pulse	500 (average)	Welding, material removal
Ar^+	0.4880 0.5145	Continuous	18	Semiconductor processing

A remarkable number of laser types are now known—nearly 2,000. They operate in oscillating wavelengths between 0.25 and 990 μm, from the ultraviolet to the submillimeter regions. Those used for processing are given in Table 13.6. Ruby, yttrium aluminum garnet, and glass lasers predominate among solid lasers and CO_2 and Ar^+ among gas lasers. The table summarizes the wavelengths, operating modes, typical power outputs, and principal applications. Many of these laser processing systems may now be found in factories. In addition, excimer lasers, a type of gas laser, have been attracting attention as the most promising of lasers for new applications such as laser chemical vapor deposition.

Let us consider the circumstances when a laser with wavelength λ, angle of divergence θ, and average power output P is focused by a lens with focal length f on a work surface, as shown in Figure 13.12. The laser energy E_0 shone on the workpiece surface declines by the depth x from the workpiece surface according to the relationship $E(x) = E_0(1 - R)e^{-\alpha x}$. The penetrating depth of laser energy is remarkably small. It is absorbed by an extremely thin layer of the workpiece surface and rapidly converted to heat. Thus, processing using lasers is thought to be carried out by the heating of the workpiece surface and a thin layer beneath it.

With each type of laser, combinations of power density and laser exposure time permit a variety of processing. When the power density is too great, a high-temperature, opaque, thermally ionized plasma develops and absorbs the laser beam, making processing impossible. At lower power densities, but greater than 10^6 W/cm^2, in an extremely brief time—on the order of 1 usec—the surface temperature reaches the vaporization point of the work material. As long as the laser exposure continues, vaporization continues, and microvolume removal is carried out, such as drilling, cut-

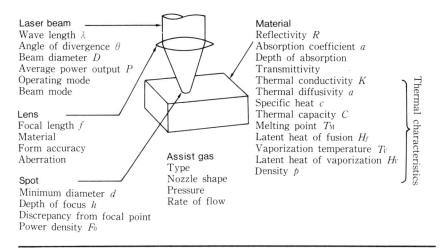

Figure 13.12. *Factors affecting laser machining.*

ting, trimming, tuning, and balancing. At a power density of 10^5 W/cm^2 or below, several microseconds are required for the surface temperature to reach the vaporization point; thus, until then, the temperature beneath the surface readily reaches the melting point of the work material. Therefore, when the laser exposure ends, welding to a certain melted depth can be effected. In addition, at a power density of 10^4 W/cm^2 or less, neither vaporization from the surface nor melting below it occur and various surface treatments can be effected by rapid cooling due to thermal conduction into the workpiece.

The major features of laser machining may be summarized as follows (Kobayashi, 1980):

1. Since the processing method uses optical systems, it is simple to set the machining position accurately, and it is possible to machine what, by conventional methods, would be an awkward spot at a difficult angle. In addition, with skillful use of the optical system, multiple processing becomes possible, especially using optical fibers with an yttrium aluminum garnet laser.

2. The power density can be varied over a wide range. By combining different power densities and laser exposure times, the same laser equipment can be used not only for removal such as drilling and cutting but also for welding and heat treating.

3. Removal processing with thermal energy at a high power density can machine hard, brittle materials including diamond, gems, and ceramics,

regardless of the hardness of the material. Micromachining is also possible.

4. Since laser processing is a noncontact machining method, it is free of the changes in cutting qualities which occur, due to tool wear, heating, and damage, in cutting and grinding processes using conventional mechanical energy. Thus, stable machining is possible. The force applied to the workpiece during laser processing is small and causes little deformation; thus it is possible to machine thin parts or places with little rib left, and no heavy jigs are needed.

5. Machining is possible with few residual strains and heat-affected zones.

6. Depending on the laser wavelength, it is possible to machine through a transparent substance.

7. Since there is no need for high voltages, as in electron beams, no X rays are produced and there is no need to work in a vacuum.

8. The operators need not be highly skilled.

9. Automation of the machining is simple and economical.

Laser processing systems that are in practical use and proving effective in actual factory applications are summarized in Table 13.7. Metals do not tend to absorb infrared light from Co_2 lasers. In contrast, ceramics readily absorb infrared light. Therefore, their efficient machining with CO_2 laser equipment, which has a relatively low power output, is possible. Thus, laser processing is particularly well suited for use with ceramics.

Sales of CO_2 laser processing systems in the United States in 1984 totaled 942 machines: 9 at 5 kW and above, 38 at 2 to 5 kW, 417 at 0.1 to 2 kW, and 478 at 100 W or less. Of these, 467 were used for metals and 475 for nonmetallic substances. The devices for nonmetallic applications can be broken down into 310 for cutting and drilling, 135 for marking, and 30 for heat treating (Hitz, 1985).

The present state of laser processing of ceramics is discussed below for each type of processing.

B. DRILLING

The starting holes for diamond dies are now almost always made by lasers. An example is given in Figure 13.13. (Kobayashi, 1982b). The starting holes for artificial rubies and sapphires used in watch bearings are drilled in the same way (Figure 13.14). The analog watch, threatened at one time by the digital watch, is produced by large-scale mass production, and demand for such bearings has grown rapidly. Figure 13.15 indicates how the machining efficiency of drilling by laser for diamond dies and watch bearings has increased over time (Kobayashi, 1979b). The efficiency gains due to the introduction of lasers have been remarkable. Recently, many cases of eliminating impurities from gem-quality diamonds by laser drilling, re-

Table 13.7. Laser Processing Methods Established as Factory Production Processes According to Frequency of Use

Processing method	Part	Material	Solid state laser			Gas laser		First application
			Ruby	YAG	Glass	CO_2	Ar	
Drilling	Wire drawing dies	Diamond	X					1966[a]
	Watch bearings	Ruby, sapphire	X	XX				1968[a]
	Heat-resistant circuit substrates	Alumina ceramics	X	XX				1975[a]
	Irrigation pipes	High-density polyethelene resin				XX		1974[b]
	Aerosol valves	Polyacetals				XX		1974[b]
	Contact lenses	Polymethyl methacrylate				XX		1974[b]
	Nipples for baby bottles	Rubber				XX		1974[b]
	Filter parts for cigarettes	Paper				XX		1966[b]
Cut-off	Jet engine turbine blades	Heat-resistant alloy	X	X	XX			1966[a]
	Die board	Plywood				XX		1970[b]
	Dress fabric	Cloth				XX		1971[b]
	Aerospace parts	Plastic composite				XX		1975[b]
	Electric, automotive, chemical	Sheet metal of various metals				XX		1975[b]
	Aerospace sheet metal	Ti alloy, Ni alloy, high tensile strength				XX		1975[b]
Scribing	Semiconductor elements	Silicon		XX				1973[c]
	Automotive, computer, aeronautic	Alumina ceramics				XX		1970[c]
Trimming	Electrical parts, resistance	Bonded thin film; coated and fired thick film		XX		X		1971[c]
Welding	Apollo rock containers	Ni-Co alloy	XX					1970[a]
	Nuclear reactor fuel rods	Stainless steel	XX					1974[a]
	Tiny relays for telephones	Gold-plated contact and spring leaf	XX					1976[a]
	Terminal contact insulated leads	Insulated leads, insulation stripped off		XX				1975[c]
	Electron gun parts	Brim (0.2-mm thick) and ribbon 0.08-mm thick		XX				1979[c]
	Coffee percolators	0.6-mm thick stainless steel				XX		1976[b]
	Ballpoint pen cartridges	Stainless steel				XX		1975[b]
	Cylindrical lead batteries	Butting together tab 5-mm thick				XX		1974[b]
Heat treatment	Power steering gear housing	Ferrite malleable cast iron				XX		1974[b]
	Diesel engine cylinder liner	For electric locomotives				XX		1978[b]

[a] Ruby laser. [b] CO_2 laser. [c] YAG laser.

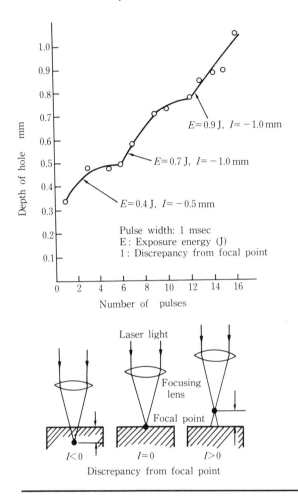

Figure 13.13. Drilling holes in diamond dies with a YAG laser.

sulting in actual improvement of the diamonds' quality as gemstones, have been reported.

Laser drilling of the alumina ceramics often used in electronic circuits has also become practicable. The hole diameter and distance between holes can be machined with high accuracy. Moreover, the cost is trivial compared with that of drilling holes with a diamond drill.

Glass transmits conventional yttrium aluminum garnet laser light, making drilling holes in glass with such lasers difficult. It has been observed that, by altering the surface roughness of the glass or by using certain types of surface coatings, yttrium aluminum garnet laser drilling becomes pos-

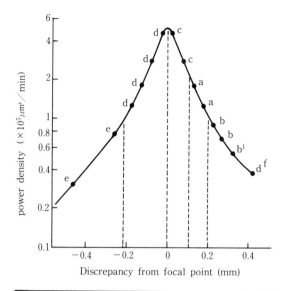

Figure 13.14. Drilling holes in ruby with a YAG laser.

sible. An example of the relationship between the cumulative exposure energy and the machined volume is presented in Figure 13.16. Finding the value for the ratio of machined volume to cumulative exposure energy (mm^3/J) indicates the characteristics of glass machined with laser light (Takezawa et al., 1984).

Also, if an aqueous solution containing transition metal ions is placed in close contact with the underside of the glass and laser light is beamed on it from the upper surface of the glass, it is possible to drill deep, small-diameter holes from the underside of the glass (Ikeno et al., 1985).

Table 13.8 presents examples of laser drilling for many types of ceramics (Yasunaga, 1984). In each case, fine holes can be drilled in a short time. Lasers with a long pulse width have a large removal volume, but hole clogging occurs readily due to leavings of the fused material. To carry out hole drilling most effectively, it is advisable to select a power output such that the boring proceeds by repeated pulses until the hole is ready.

C. CUTTING

At present the most widely used method of cutting ceramics uses a diamond grinding wheel. This method, however, is limited to cutting straight lines—contoured cutting is not yet possible—and there are still many unsolved problems, such as a diamond wheel dressing method. The use of

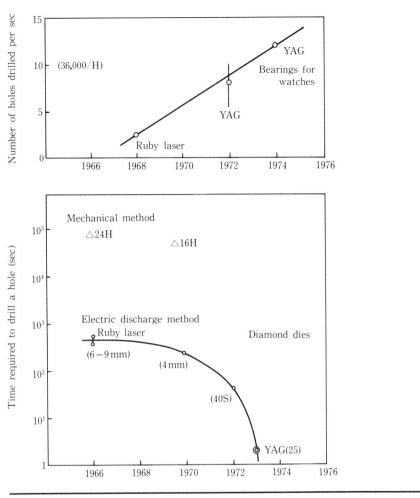

Figure 13.15. Increase in efficiency with laser drilling.

Table 13.8. Examples of Laser Hole Drilling

Part, material	Thickness (mm)	Hole diameter (mm)	Laser used	Laser power output	Number of pulses	Drilling time (sec)
Diamond dies	1	0.05	YAG	0.5–1.0 J	≈20	2
Ruby bearings	0.3	0.06	YAG	40 kW	40	0.1
Ferrite	1	0.1	YAG	0.5–1.0 J	≈10	1
Aluminum ceramics	3	0.2	YAG	1–2 J	≈50	5
Aluminum ceramics	3	0.2	YAG	1–2 J		5
	3.2	0.25		1.4 J		8
	0.6	0.5	CO_2	75 W		0.2
	0.7	0.25	CO_2	250 W		1
Fused quartz	3	0.2	CO_2	600 W		3
	5.8	1.6	CO_2	100 W		3

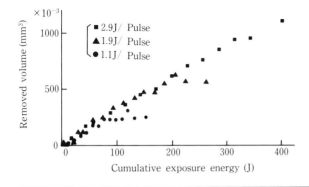

Figure 13.16. *Relationship between cumulative exposure energy and removed volume.*

lasers instead of a diamond wheel increases not only permits cutting in complex shapes but also increases the cutting speed and lowers costs.

Sheets of materials in the SiO_2 family (fused quartz, for instance) are cut with a CO_2 laser; quite high cutting speeds are possible. Figure 13.17 shows the results from cutting a rotating fused quartz cylinder (external diameter 18 mm, walls 1.5-mm thick) with a 450 W CO_2 laser, blowing N_2 gas on it at a rate of 30 L/mm during the cutting. As the rotational frequency of the sample rises, its effect on the cutting speed declines. The cutting surface is in all cases unusually smooth. Since burrs and microcracks are hardly to be discerned, post-cutting finishing is not necessary.

The CO_2 laser can even be used to cut extremely hard ceramics, such as Si_3N_4. The efficiency is higher than when diamond wheels are used. Moreover, cutting curved lines is also possible. Thus, this cutting method offers great promise. Table 13.9 presents a summary of examples of cutting inorganic, nonmetallic materials with CO_2 lasers (Kobayashi, 1983b; Yasunaga, 1985).

D. SCRIBING

Scribing refers to the cutting of grooves in semiconductor wafers and ceramic substrates to divide them into pellet form. In the past, scribing was performed mechanically, using a diamond point. With diamond scribing, however, the grooves are shallow (on the order of 10 μm) but the creation of microcracks is unavoidable. Thus, a high proportion of the materials are damaged on being broken into pellets, and reaching a yield of 90% or better has been regarded as difficult. Research on the use of laser light as a scriber became lively around 1970. As its superiority became apparent, this laser application was put to practical use. At present, replacement of diamond scribers by lasers scribers is proceeding steadily.

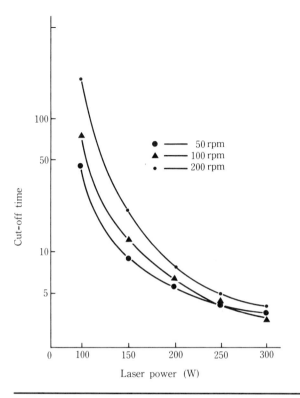

Figure 13.17. Cutting off a quartz tube with a CO_2 laser.

Table 13.9. Results of Cut-off of Inorganic, Nonmetallic Materials by CO_2 Laser

Material	Sheet thickness (mm)	Cut-off speed (mm/min)	Kerf (mm)	Laser power (kw)	Laser manufacturer[a]
Ordinary glass	9.5	1,500	1.5	20	1
Pyrex glass	2.2	500		0.3	
Quartz glass	1.9	600	0.2	0.3	2
	1	2,500		0.5	
Fused quartz	9.5	130		1	3
Ceramic tile	6.4	500		0.85	
Alumina ceramics	0.6	2,300		0.5	
Asbestos cement	5	600	0.1	0.3	2
Concrete	38	50	0.3	8	1

[a] 1—AVCO, U.S.A.; 2—Messer Griesheim, West Germany; 3—Spectra Physics, U.S.A.

13. Precision Machining Methods for Ceramics

The laser scribers have a number of advantages:

1. Compared with diamond scribers, the cut is uniform, the groove width is narrow (a minimum of 15 to 25 μm), but a large groove depth (50 to 200 μm) can be machined. Thus, the yield when the material is broken into pellets has improved to 99% or better.
2. Reciprocating cutting is possible. Moreover, since there is no tool wear, the machining efficiency is remarkably high.
3. The process is easy to automate; it does not require skilled operators.

At present, repeating pulse yttrium aluminum garnet or CO_2 lasers of kilohertz or greater repetition are used in laser scribing. Yttrium aluminum garnet lasers are principally used to scribe Si wafers, while CO_2 lasers scribe ceramics. An example of the machining characteristics of the scribing of Si wafers with an yttrium aluminum garnet laser is given in Figure 13.18 (Kobayashi, 1982b).

Many 50 to 500 W CO_2 lasers are used in scribing ceramics. An example of scribing an aluminum sheet 0.635-mm thick at a machining speed of 200 mm/sec by repeated pulses of 1.2 kHz with an average power output of 50 W, and a pulse width of 100 msec has been reported (Sugijima et al., 1984). Results show that the scribing produces a row of conic holes.

A problem in laser scribing is the adhesion of scattered particles. By selecting the appropriate conditions, including spraying with an inert gas and keeping the machining width narrow and the removal volume to a minimum, scribing is possible without needing a washing process. As a general characteristic, good results are obtained if the machining speed (mm/sec) is more than six times the pulse frequency (kHz) (Suzuki et al., 1972).

Many advantages have been recognized in using high-output yttrium aluminum garnet lasers instead of CO_2 lasers in scribing (Laser Institute of America, 1979). A CO_2 laser leaves a large melted section in the area around the spot, so that the cross-section width affected by the heat is 0.2 mm. In contrast, with yttrium aluminum garnet lasers that width is only 0.05 mm. In addition, comparisons of the flexural strength of ceramics after scribing shows that those scribed by a Q-switched yttrium aluminum garnet laser are 45% stronger than those scribed by CO_2 laser or pulse yttrium aluminum garnet laser; they are also 13% stronger than ceramics scribed with diamond.

To make the liquid-crystal displays used in the watch and electronics industry, it is necessary to scribe the individual displays accurately from a large glass substrate. Scribing with a CO_2 pulse laser by the noncontact system has become practical. In some types of glass, cold flow may occur after exposure to lasers. Thus, it is preferable to divide the glass substrate immediately after laser scribing.

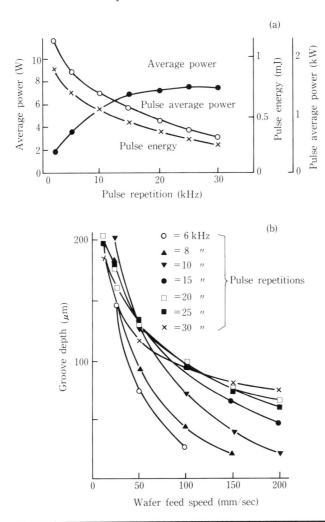

Figure 13.18. Scribing characteristics of Si wafers, using a YAG laser. (a) shows the output characteristics of a YAG laser scriber and (b) shows the relationship between wafer feed speed and groove depth.

As sapphire has become increasingly popular as a semiconductor substrate, its scribing has put laser scribing in the limelight.

Just as lasers can drill holes in diamonds, they are frequently used to scribe diamonds. A Q-switched yttrium aluminum garnet laser is used. This laser function is excellent for processing gem-quality diamonds, since it can cut in a complex two-dimensional shape with a small cutting width, without influencing the crystal's orientation.

E. LASER-ASSISTED ETCHING

Laser light is not only used for direct heat removal machining. A recently active field of research concerns laser-assisted etching technology, which uses the effects of pyrolysis and photolysis in a chemical solution or a reactive gas atmosphere. It is possible, using a short wavelength laser, to make minute grooves and three-dimensional patterns at will. In addition, laser-assisted etching is also effective as a resistless etching method. It offers noteworthy new directions for lithographic techniques as well.

The latest research report comes from von Gutfeld and Hadgson at IBM's Thomas J. Watson Research Center (von Gutfeld and Hadgson, 1982). They placed silicon or alumina/TiC (70:30) ceramic in a 2 to 18 mol KOH solution in a case made of fused quartz and exposed it to Ar laser light. For a silicon (111) surface, it is possible, using a 15 W laser with 10^7 W/cm^2, to bore a hole through a wafer 25 μm thick at a speed of 15 μm/sec. For ceramics, reported results indicate an instantaneous etching speed of 200 μm/sec at 10^6 W/cm^2.

The instantaneous removal volume changes depending upon the pulse interval and the power density. To make the spot smaller, the beam is first expanded, then condensed.

When ceramics are machined, fine black particles fly from the sample. After the laser exposure ends, these particles are dissolved in a few seconds by the KOH, and the KOH solution again becomes transparent. With a ceramic sample 0.9-mm thick, an approximately one-second exposure from an Ar^+ laser beam with output of 1.5 W will bore a blind hole about 0.1 mm in diameter.

Results found for the relationship between laser output and volume removal rate for these materials are presented in Figure 13.19. For silicon, machining with less than 3.3 W was not possible. For ceramics, 1 W or more was needed. The difference between these threshold values is believed to lie in the differences in the thermal constants of these substances, that is, their thermal conductivities and melting temperatures, and the difference in the power density due to differences in each laser spot.

In any case, these are new attempts indicating the possibility of machining microholes by laser.

Several research results have been reported recently in Japan (Koyabu, 1984; 1985; Mihashi, 1985). Figure 13.20 gives an example of yttrium aluminum garnet laser etching equipment. The higher the KOH concentration and the greater the laser power, the greater the machined volume. In addition, the ratio of groove depth to groove width occurs in the following ascending order of laser types: CW < single < Q-switched < Q-switched plus single (Koyabu, 1984). If an Ar laser, which has a short wavelength (about 0.5 μm) is used, finer machining is possible (see Figure 13.21). If the laser power is reduced, a line width of 1 to 2 μm can be attained (Mihashi, 1985).

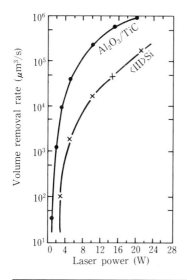

Figure 13.19. Influence of laser power output on volume removal rate. From van Gutfeld and Hadgson (1982).

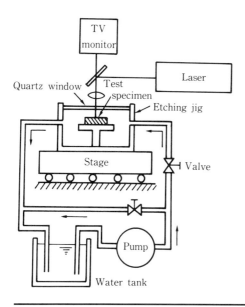

Figure 13.20. Structure of a laser-assisted chemical etching device. From Koyabu et al. (1984).

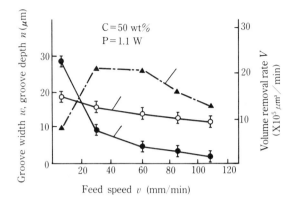

Figure 13.21. Feed speed of the workpiece and characteristics of the grooves machined. From Mihashi (1985).

Dry etching with a reactive gas has also been attempted using transverse electric atmospheric CO_2 lasers, Ar lasers, and excimer lasers. Figure 13.22 presents an XeCl excimer laser (0.308 μm, 10-nsec pulse width, pulse repetition of 80 pps, average power density 3 W/cm²) shone on a polycrystalline Si surface after passing through a pattern formed of aluminum on a quartz mask. This brought about resistless etching by causing an etching

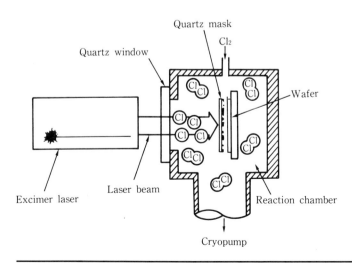

Figure 13.22. Resistless etching of Si wafers by excimer laser.

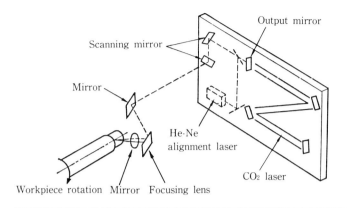

Figure 13.23. Laser machining. From Copley et al. (1978).

reaction between the laser-excited Si and the Cl* radical due to photolysis of the Cl_2. This resistless laser etching technique promises to reduce processing time considerably.

F. LASER MACHINING

As Figure 13.23 shows, laser machining is an attempt to use-laser-focused light to machine grooves directly on the surface of ceramics. Results have been published reporting machining grooves in Si_3N_4 and SiC, for instance, using 450 W or 1.4 kW CO_2 lasers (Copley et al., 1979a, 1979b, 1981).

VI. OTHER MACHINING METHODS

In this section we briefly discuss machining methods for ceramics other than those discussed above.

A. ULTRASONIC MACHINING

Ultrasonic machining has been in widespread use for some time now for boring holes, cutting, and carving glass, ferrite, quartz, and gemstones. It can be applied to any hard, brittle material, whether or not it is electrically conductive.

In this machining method, the ultrasonic oscillating tool is pressed against the workpiece at a set load. A slurry containing abrasive grains is provided in the small gap at the tip of the tool. The abrasive grains are driven percussively into the workpiece, causing brittle fractures to accumulate, and the shape of the tool tip is transferred to the workpiece.

Therefore, ultrasonic machining is suited for hole boring and cutting ceramics, which are low in tensile strength. Since there is no need for the tool to rotate, the technique can be used for boring and carving specially shaped holes, not just round ones.

Figure 13.24 shows the relationship between the applied pressure and the horizontal feed rate on the removal volume rate when hot-pressed Si_3N_4 is machined ultrasonically with a slurry containing B_4C No. 280 (Ishiwata, 1985). In the illustration, the data are for machining ring grooves with a rotating pipe-shaped oscillating tool. Under optimal conditions, a removal volume rate of about 70 mm^3/min was obtained.

There are relatively few data on the ultrasonic machining of ceramics other than alumina and ferrite. There is, however, little wear on the tool and little deformation, so that machining can proceed in a stable state. The machining speed is also high. Thus, ultrasonic machining is used quite widely in factories (Fuji, 1980). There are also examples of the practical use of ultrasonic hole boring, core drilling, lapping, and grinding (Ishiwata, 1985).

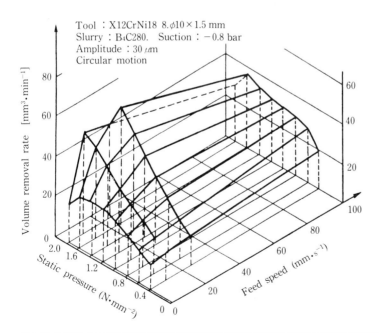

Figure 13.24. Relationship of volume removal rate, static pressure, and feed speed in the ultrasonic machining of pressed Si_3N_4.

B. CUT-OFF METHODS

As Table 13.2 suggests, there are many troublesome aspects of cutting ceramics. Cutting methods for nonmetallic materials are summarized in Tables 13.10, 13.11, and 13.12, (Kobayashi, 1985). As can be seen, the methods in use for cutting ceramics include grinding-cutting by ordinary grinding wheels, cut-off by diamond wheels, lapping by a rotating disk, lapping by the multiblade method, lapping by the multiwire method, cut-off by blasting, cut-off by ultrasonics, cut-off by wire electric discharge machining, and laser cut-off.

A comparatively large literature exists on the various cut-off methods using grinding wheels and diamond wheels for ceramics (Ishida, 1980; Komine, 1984; Kobayashi, 1985b; Ōshita, 1982). The selection of a diamond wheel or the development of new types of wheels is regarded as necessary, as is the development of dressing techniques for their use. There

Table 13.10. Cut-off Methods for Single-Crystal Materials

Material	Energy source	Cut-off method	Cut-off principle
Semiconductors Si	Mechanical	Abrasive machining	
		Bonded abrasives	
		Diamond wheel	Dicing, external type
			Internal type
			Diamond band saw
		Loose abrasives	
		Lapping	Multiblade, multiwire
		Ultrasonic	Ultrasonic vibration
		Shearing	
		Press	Lateral pressure
	Electrical	Electrophysical machining	Scribing
		Shearing (specialized equipment)	
		Laser machining	Scribing
Compounds	Mechanical	Abrasive machining	
		Bonded abrasives	
		Diamond wheel	Dicing, external type
			Internal type
			Diamond band saw
		Loose abrasives	
		Lapping	Multiwire
		Other	
		Fluid blasting	Cleavage
	Electrical	Electrophysical (specialized equipment)	
		Laser machining	Scribing

Table 13.10. (Continued)

Material	Energy source	Cut-off method	Cut-off principle
Dielectric, piezoelectric Quartz	Mechanical	Abrasive machining Bonded abrasives	
		Diamond wheel	Slicing, external type Internal type Diamond band saw
		Loose abrasives	
		Lapping	Multiblade, multiwire
		Ultrasonic	Ultrasonic vibration
		Fluid blasting	Liquid honing Cleavage
	Electrical	Electrophysical (specialized equipment)	
		Laser machining	Continuous
Other	Mechanical	Abrasive machining Bonded abrasives	
		Diamond wheel	Slicing, external type Internal type
		Loose abrasives	
		Lapping	Multiwire
Ferrite	Mechanical	Abrasive machining Bonded abrasives	
		Diamond wheel	Slicing, external type Internal type Diamond band saw
		Loose abrasives	
		Lapping	Multiblade
		Ultrasonic	Ultrasonic vibration
Optical elements	Mechanical	Abrasive machining Bonded abrasives	
		Diamond wheel	Slicing, external type
Gemstones	Mechanical	Abrasive machining Bonded abrasives	
		Diamond wheel	Slicing, external type

are a number of major issues concerning the cutting off of ceramics: wheel life, variation in wheel shape, cut-off accuracy, deterioration of the cut-off surface, chipping at the cut-off edge, and productivity. Obviously, there is a need for the development of cut-off machine tools suited for use with ceramics.

A cut-off method using hydrostatic pressure has also been developed (Satō, 1976). Its basic principles are as shown in Figure 13.25. In lateral pressure cutting, hydrostatic pressure is applied to the exterior of a cylindrical workpiece, both ends of which are free. When a certain pressure is

Table 13.11. Cut-off Method for Ceramics

Material	Energy source	Cut-off method	Cut-off principle
Alumina	Mechanical	Abrasive machining	
		Bonded abrasives	
		Diamond wheel	Slicing, external type
			Dicing, external type
		Loose abrasives	
		Ultrasonic	Ultrasonic vibration
	Electrical	Electrophysical machining	
		Shearing (specialized equipment)	
		Laser machining	Continuous, scribing
Ferrite	Mechanical	Abrasive machining	
		Bonded abrasives	
		Diamond wheel	Slicing, external type
			Dicing, external type
			Diamond band saw
		Loose abrasives	
		Lapping	Rotary disk, multiblade, multiwire
	Electrical	Electrophysical machining	
		Shearing (specialized equipment)	
		Laser machining	Scribing
Nitrides	Mechanical	Abrasive machining	
		Bonded abrasives	
		Diamond wheel	Slicing, external type
		Loose abrasives	
		Ultrasonic	Ultrasonic vibration
	Electrical	Electrophysical machining	
		Shearing (specialized equipment)	
		Electric discharge machining	Wire electric discharge
		Laser machining	Continuous, scribing
Carbides	Mechanical	Abrasive machining	
		Bonded abrasives	
		Diamond wheel	Slicing, external type
		Loose abrasives	
		Ultrasonic	Ultrasonic vibration
	Electrical	Electrophysical machining	
		Shearing (specialized equipment)	
		Laser machining	Continuous, scribing

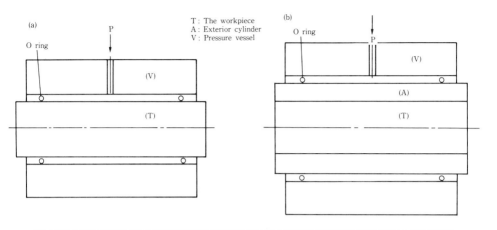

Figure 13.25. Principle of the cut-off method using hydrostatic pressure. (a) shows the lateral pressure cutting method, and (b) shows the disking method, From Satō (1976).

Table 13.12. Cut-off Method for Other Materials

Material	Energy source	Cut-off method	Cut-off principle

Stone and concrete	Mechanical	Abrasive machining
		  Bonded abrasives
		    Grinding wheel
		    Diamond wheel
		  Loose abrasives
		    Ultrasonic
			
			Abrasive
			Slicing, external type
			Diamond band saw
			Ultrasonic vibration

Glass	Mechanical	Abrasive machining
		  Bonded abrasives
		    Grinding wheel
		    Diamond wheel
		
		    Lapping
		  Loose abrasives
		    Ultrasonic
		Shearing (press)
			
			Abrasive
			Slicing, external type
			Diamond band saw
			Multiblade, multiwire
			
			Ultrasonic vibration
			Lateral pressure

	Electrical	Electrophysical machining
		  Shearing (specialized equipment
		  Laser machining
			
			Continuous

(continued)

Table 13.12. (Continued)

Material	Energy source	Cut-off method	Cut-off principle

Cermet	Mechanical	Abrasive machining
		  Bonded abrasives
		    Diamond wheel
		
		Shearing
		  Shearing (specialized
		    equipment)
			
			Slicing, external type
			Dicing, external type
			Diamond band saw
			
			Guillotine shear, slitter

	Electrical	Electrophysical machining
		  Shearing (specialized
		    equipment)
		  Laser machining
			
			Scribing

Plastic	Mechanical	Cutting
		  Blades
		    Cutting tools
		    Saws
		Abrasive machining
		  Bonded abrasives
		    Grinding wheel
		Other
		  Fluid blasting
		  Shearing (press)
		  Shearing (specialized
		    equipment)
			
			Bite, cutter
			Circular, band
			
			Abrasive
			
			Liquid honing
			Precision punching
			
			Guillotine shear, slitter

	Thermal	Thermal machining Heat of friction	 Electric heating wire
	Electrical	Electrophysical machining Laser machining	 Continuous

Composite materials	Mechanical	Cutting
		  Blades
		    Cutting tools
		    Saws
		Abrasive machining
		  Bonded abrasives
		    Grinding wheel
		    Diamond wheel
			
			Bite, cutter
			Circular, band
			
			Abrasive
			Slicing, external type

	Electrical	Electrophysical machining
		  Shearing (specialized
		    equipment)
		  Laser machining
			
			Continuous

reached at that horizontal section, the workpiece develops a tension fracture. In the disking method, an exterior cylinder in which an entry cut has been made covers the workpiece. Thus multiple tubes of the same thickness can be cut in the same way. It has been reported that mirror surface cut-off of Pyrex glass, quartz, alumina ceramics, marble, ruby, sapphire, and other hard, brittle materials between 1.5 and 3 mm thick has been achieved. No cutting tool is needed, so that tool wear and changes in cut-off qualities are not an issue. All that is required are a pressure vessel and a pressurizing pump. The cut-off is instantaneous. Thus, hydrostatic pressure cutting has many advantages. It appears, however, that it is difficult to produce thin wafers (1-mm or less thick) with this method.

C. ELECTRICAL APPLICATIONS MACHINING AND COMPLEX MACHINING

Electrical applications machining includes the machining processes using electrochemical, electrophysical, and optical energy, as presented in Table 13.1. Machining methods utilizing optical energy were covered in Section IV of this chapter. Here let us sketch the more important of the electrochemical and electrophysical methods.

1. Electrical Discharge Machining (Kubota, 1984)

Since ceramics are usually electrically nonconducting or semiconducting, in machining them the effect of electrical resistance is great, necessitating the use of special methods. An electrical discharge method of boring holes in diamonds was developed in the mid-1950s by Hoh and Kurafuji. In this method, a diamond is immersed in an electrolytic solution and electricity is discharged between the metal needle-form electrode and the electrolytic solution. The effect is to bore a hole in the diamond at the bottom surface of the solution. Based on that concept, electrolytic electric discharge wire-cut machining was tested using a $2N$ KOH solution (Miyazaki et al., 1983). Compared with wire-cut machining of metals, the machining speed when the method is used with Al_2O_3 or Si_3N_4 is reported to be strikingly slower. Electric discharge machining of SiC, utilizing the material's semiconducting qualities, is being researched in many places.

The material produced when TiN is mixed with Si_3N_4 and hot-pressed has distinctive features as a semiconducting ceramic. It is readily processed by electric discharge machining. Its machining characteristics under electric discharge machining are quite similar to those of copper.

2. Electric Discharge Complex Grinding

A grinding wheel as illustrated in Figure 13.26 has been developed which is designed to suit all types of ceramics (Kuromatsu, 1984). With a 0.5 to 1%

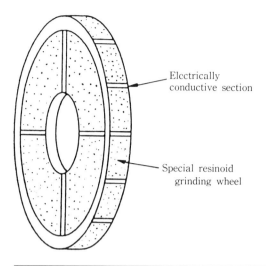

Figure 13.26. Structure of MEEC grinding wheel. From Kuromatsu (1984).

$NaNO_3$ solution, grinding proceeds while weak electric discharges are produced. The method is demonstrating its machining ability by shaping Sendust (an alloy of Fe, Al, and Si), which is attracting attention as a material for magnetic heads (Takizawa, 1983; *Nikkei Mechanical*, 1984).

3. Electrolytic Complex Grinding

These methods combine, in addition to electrolytic grinding and mechanical grinding, the anode oxidation phenomenon. They are used to machine Sendust and other materials (Ishiyama, 1983).

4. Ion Beam Machining

Aspherical glass lenses, the tips of diamond indenters, piezoelectric elements, and quartz oscillators are machined by ion beam machining (Taniguchi, 1980).

REFERENCES

Arikado T. et al. (1984). *Kōkagaku hannō no handōtai purosesu e no ōyō* [Applications towards optochemical reaction machining of semiconductors]. *Rēza kenkyū*, 12–7, 348–356.

Brehm, R. K. van Dun; Teunissen; J. C. G. and J. Haisma. (1979). Transparent single-point turning of optical glass. *Precision Engineering*, 1, 207–213.

Bryan, J. B. (1979). Construction of an ultraprecision 84-inch diamond turning machine. *Precision Engineering,* 1.
Bryan, J. B. (1981). The power of deterministic thinking in machine tool accuracy. Paper presented at First International Precision Engineering Seminar, May, 26–28.
Bryan, J. B.; Clouser, R.; and Holland, E. (1967). How to measure spindle performance and put the results to work. *American Machinist,* 12, 142–166.
Carter, D. (1982). An 84-inch diameter horizontal (spindle diamond turning machine (D. T. M. 3). Paper presented at a Society of Manufacturing Engineering Seminar, June 8.
Ceramics Japan. (1983). *Ankēto chōsa—seramikku seihin oyobi hanseihin no jokyo kakō no genjō* [Questionnaire: The state of removal processing of ceramic products and semi-finished products]. *Ceramics Japan,* 18, 506–515.
Copley, S. M., Bass, M., and Wallace, R. G. (1978). Shaping silicon compound ceramics with a continuous carbon dioxide laser. Proceedings, 2nd International symposium on Ceramic Machining and Finishing.
Daitō, S. (1985). *Kikai to kōgu,* 4, 76–86.
Donaldson, B. (1982). A 60-inch diameter vertical spindle diamond turning machine (LODTM). Paper presented at a Society of Manufacturing Engineering Seminar, June 8.
Fuji, Y. (1980). *Seramikkusu no chōonpa kako* [Ultrasound machining of ceramics]. *Kogyō zairyō,* 28, 39–42.
Groenou, A. B. et al. (1978/79). *Philips Technical Review,* 4/5.
Hitz, C. B. (1985). Marketplace for materials-processing lasers. *Lasers and Applications,* 59–61.
Ikeno, J. et al. (1985). *YAG rēzakō o garasu ni shoshashita toki no shō genshō* [Various phenomena that occur when YAG laser light is shined on glass]. Paper presented at the Spring meeting of the Japan Society of Precision Engineering.
Imanaka, O. (1967). *Muki zairyō no seimitsu kakōhō* [Precision machining methods for inorganic materials], *Kikai no kenkyū,* 19, 61–67.
Ishida, Y. (1980). *Seramikkusu no kensaku kako* [Grinding ceramics]. *Kōgyō zairyō,* 28, 29–33.
Ishiwata, S. (1985). *Chōonpa kakō* [Ultrasound machining]. *Kogyō zairyō,* 33, 111–117.
Ishiyama, M. (1983). EMG machining of brittle materials. *Kikai to kōgu,* Vol. 27, No. 94–98.
Itō, K. (1984). *Seramikkusu kakō ni okeru kensakuban no arikata* [An ideal grinder for ceramics]. *Kikai gijutsu,* 32, 36–40.
Itō, M. (1981). *Nansakusei seramikkusu no seimitsu kako* [Precision machining of ceramics that present difficulties in removal machining]. Paper presented at the Ceramics Society of Japan conference on high-temperature materials technology, October.
Itō, M. (1983). *Chikkabutsu kei, tankabutsu kei seizōyō seramikkusu no seimitsu kensaku* [Precision grinding of nitride and carbide ceramics for precision applications]. *Ceramics Japan,* 18, 479–485.
Iwata, S. (1980). *Kaisakusei kesshoka garasu: makōru no sekkaku kakō* [Readily machinable crystallized glass: Cutting Macor]. *Kōgyō zairyō,* 28, 55–59.
Juchem, H. O. et al. (1979). Precision machining of non-metalic materials—Some practical guidelines. *Industrial Diamond Review,* 2, 43–51.

Karaki, T. et al. (1978). Polishing characteristics of GGG single crystals using acid solutions. *Journal of the Japan Society of Precision Engineering,* 44, 333-339.

Kasai, T. (1978a). *LiTaO$_3$ tan kesshō no rappingu tokusei* [Lapping characteristics of LiTaO$_3$ single crystals]. *Journal of the Japan Society of Precision Engineering,* 4411, 1360-1362.

Kasai, T. et al. (1978b). *GGG tan kesshō no sansei yōeki ni yoru mekanokemikaru porishingu tokusei* [Characteristics of mechanochemical polishing of GGG single crystals with acidic solutions]. *Journal of the Japan Society of Precision Engineering,* 443, 341-348.

Kobayashi, A. et al. (1954). *Pairekkusu garasu no so kensaku hō no kenkyū* [Studies of rough grinding methods for Pyrex glass]. *Journal of the Japan Society of Precision Engineering,* 206, 208-211.

Kobayashi, A. (1970). Recent development in diamond wheel grinding. *Annals of the CIRP,* 18, 31-37.

Kobayashi, A. (1978). High-precision cutting with a new ultra precision spindle. *Annals of the CIRP,* 27.

Kobayashi, A. (1979a). *Koseido kūki jikuju kaitenjiku no kaihatsu no rekishi to sono ōyōrei* [History of the development of the high precision air bearing spindle and sample applications]. *Journal of the Japan Society of Precision Engineering,* 45, 1159-1162.

Kobayashi, A. (1979b). Development of application laser techniques *Kikai to kōgu,* Vol. 23, no. 12, 25-41.

Kobayashi, A. (1980). Precise machining with newly developed high precision air bearing. *SME Technical Paper.* MR 80-214.

Kobayashi, A. (1982a). *Chōseimitsu daiyamondo sekkaku kakō gijutsu* [Ultraprecision diamond cutting technique]. *Journal of the Japan Society of Mechanical Engineers,* 85, 241-248.

Kobayashi, A. (1983a). *Serammikkusu no rēza kakō—hatten keika to shōrai dōkō* [Laser machining of ceramics—The course of development and future trends]. *Ceramics Japan,* 18, 498-505.

Kobayashi, A. (1983b). Recent development of ultra-precision diamond cutting machine in Japan. *Bulletin of the Japan Society of Precision Engineering,* 172, 73-77.

Kobayashi, A. (1985a). Ultra precision diamond cutting machine. *Sekkei seizu,* 20, 94-.

Kobayashi, A. (1985b). *Saishin setsudan gijutsu sōro* [Discussion of the latest cut off techniques]. Sangyō gijutsu sābisu sentā.

Komine, S. (1984). *Daiyamondo toishi ni yoru seramikkusu no kensaku kako* [Grinding ceramics with a diamond wheel]. *Kikai gijutsu,* 32, 41-46

Koyabu, K. et al. (1984). *Rēza ashisuto kemikaru etchingu* [Laser-assisted chemical etching]. Paper presented at the Fall meeting of the Japan Society of Precision Engineering.

Koyabu, K. et al. (1985). *Rēza ashisuto kemikaru etchingu* [Laser-assisted chemical etching]. Paper presented at the Spring meeting of the Japan Society of Precision Engineering.

Kubota, M. (1984). *Seramikkusu no hōden kako oyobi hōden fukugo kensaku* [Electric discharge machining of ceramics and multiple electric discharge grinding]. *Kikai gijutsu,* 32, 76-80.

13. Precision Machining Methods for Ceramics 311

Kuromatsu, A. (1984). *Seramikkusu no denkai, hōden kensaku kako* [Electric discharge grinding of ceramics]. Paper presented at the January meeting of the Japan Society of Grinding Engineers.

Laser Institute of America. (1979). *Guide for Material Processing by Lasers.* Baltimore: The Paul M. Harrod Co.

McKeown, P. A. (1982). Design and Construction of a 5-inch diameter vertical spindle diamond turning machine. *SME Technical Paper,* MR82-931.

Mihashi, M. (1985). *Rēza yūki etchingu ni seramikku no bisai kakō* [Fine machining of ceramics by laser etching]. Paper presented at the Spring meeting of the Japan Society of Precision Engineering.

Miyashita, M. and Yoshioka, J. (1980). *Kōsaku kikai no undōgaku — kakō seido no tachiba kara* [The kinetics of machine tools in terms of the machining system]. *Journal of the Japan Society of Mechanical Engineers,* 83, 1152–1158.

Miyazaki M. et al. (1983). *Denkai hōden ni yoru seramikkusu no setsudan kako* [Electric discharge cut-off machining of ceramics]. Paper presented at the Fall meeting of the Japan Society of Precision Engineering.

Mochizuki M. et al. (1983). Precision cutting of brittle materials by diamond wheels. *Kikai to kōgu,* Vol. 27, no. 8, 66–75.

Mori, Y. (1980). Elastic emission machining to sono hyomen [Elastic emission machining and surfaces]. *Journal of the Japan Society of Precision Engineering,* 466, 659–664.

Moritomo, S. et al. (1983). *Kōzei zairyō no kikai kakōki ni yōkyūsareru tokusei* [Characteristics required for the machining of hard materials]. *Ceramics Japan,* 18, 473–478.

Murai, T. (1980). *Kikai kakō kanō mika seramikku* [Machinable mica ceramics]. *Nikkei Mechanical,* September 1, 58–63.

Nakagawa, T. et al. (1984). Paper presented at the 5th International Conference on Production Engineering, Tokyo, July.

Nakai, T. (1984). *Seramikkusu no sekkaku kakō* [Cutting ceramics]. *Kikai gijutsu,* Vol. 32, no. 8, 71–75.

Nakajima, T. (1985). Machining of Ceramics. *Kikai to kōgu,* Vol. 29, no. 4, 43–50.

Nakao, H. et al. (1985). Machining of advanced ceramics by diamond wheels. *Kikai to kōgu,* Vol. 29, no. 4, 59–63.

Narutaki, N. et al. (1984). *Shomen furaisu sessaku toki no kōgu sonshō ni oyobosu fun'iki no eikyyō* [Atmospheric effects on tool damage in face milling]. Paper presented at the Spring meeting of the Japan Society of Precision Engineering.

Nihon kikai kōgyō rengōkai. (1984). *Fain seramikkusu in kansur yūsā nīzu chōsa hōkokusho* [Report of a survey of user needs with respect to fine ceramics]. Tokyo: Nihon kikai kōgyō rengōkai.

Nikkei Mechanical (1984). Newly developed YAG laser drilling method for glass. August 13.

Niwa, T. et al. (1982). *Jiki disuku kiban no chōseimitsu kako* [Ultra-precision machining of magnetic disk substrates]. *Journal of the Japan Society of Precision Engineering* 48, 516–520.

Ogyūda, Y. et al. (1984). *Chūtetsu bondo daiyamondo toishi ni yoru seramikkusu no kakō* [Machining ceramics with a diamond-cast iron bonded grindstone]. *Kikai gijutsu,* 32, 51–56.

Ōhara, Y. (1984). *Daiyamondo toishi ni yoru seramikkusu no setsudan kako* [Cut-off machining of ceramics with diamond grindstones]. *Kikai gijutsu,* 32, 47–50.

Okutomi, M. et al. (1979). [Hydration polishing" of sapphire—A novel finishing technique using hydration phenomenon. *Bulletin of the Japan Society of Precision Engineering*, 4, 156–162.

Ōshita, H. (1982). *Seramikkusu no setsudan kakō* [Cut-off machining of ceramics]. *Kikai gijutsu*, 30, 39–44.

Pardue, B. (1982). A 60-inch diameter horizontal spindle diamond turning machine (POMA). Paper presented at a Society of Manufacturing Engineering Seminar, June 8.

Saito, T. T. (1978). Diamond turning of optics: The past, the present, and the exciting future. *Optical Engineering*, 17.

Satō, K. (1976). *Sokuatsu ni yoru zeisei zairyō no setsudan* [Cut-off of brittle substances by lateral pressure]. *Bulletin of the Japan Society of Mechanical Engineers*. 42, 2250–2257.

Satō, K. (1979). *Disukingu ni kansuru kiso kenkyū—Dai ippō* [Basic research on disking—First report]. *Bulletin of the Japan Society of Mechanical Engineers*, No. 790-17, 127–129.

Sugijma, N. et al. (1984). *Seramikkusu no rēza kakō* [Laser machining of ceramics]. *Kikai gijutsu*, 32, 66–70.

Sugita, T. (1978). *Hyomen no muhizumi: seijō chōseimitsu kakō* [Surfaces without warping: Pure ultra-precision machining]. *Kikai gijutsu*, 26, 95–98.

Sumiya, M. et al. (1981). Machining of mirror-like surfaces—Diamond cutting with new developed high-precision air bearing spindle. *Bulletin of the Japan Society of Precision Engineering*, 16.

Suzuki, M. (1980). *Toyoda Koki Technical Review*, 21–22, 12–19.

Suzuki, M. et al. (1972). *Mirikon ueha no rēza sukuraibu* [Laser scribing Millikan wafers]. Paper presented at the Fall meeting of the Japan Society of Precision Engineering.

Tajima, T. et al. (1981). *Chōseimitsu shomen senban* [Ultra-precision face lathe]. *Journal of the Japan Society of Mechanical Engineers*. 84, 1253–1258.

Takeyama, H. (1980). *Shin kōgu zairyō to takessho daiyamondo kōgu* [New tool materials and polycrystalline diamond tools]. Paper presented at the Japan Society of Precision Engineering, February 15.

Takezawa, S. et al. (1984). *YAG rēzakō o garasu ni shosha shita toki no sho genshō* [Various phenomena that occur when YAG laser light is shined on glass]. Paper presented at the Engineering.

Takizawa, T. (1983). *Kikai to kōgu*, Vol. 29, no. 87–93.

Taniguchi, N. (1980). *Seramikkusu no ion bīmu Kakō* [Ion beam machining of ceramics]. *Kōgyō zairyō*, 28, 48–54.

Tanaka, K. (1982). *Metalworking Engineering and Marketing*, 1.

Thompson, D. C. et al. (1982). Development of an inexpensive high accuracy diamond turning machine. *Precision Engineering*, 4, 13–77.

Tsujigō, Y. (1982). *Seramikkusu no kensaku kako* [Grinding ceramics]. *Kikai gijutsu*, 30, 45–50.

Tsujigō, Y. et al. (1983). *Kikai to kōgu*, 57.

Usuki, M. et al. (1982). *Fujikoshi Engineering Review*, 38.

Usuki, M. et al. (1985). *Kikai to kōgu*, Vol. 29, no. 4, 95–101.

von Gutfeld, R. J. and Hadgson, R. T. (1982). Laser-enhanced etching in KOH. *Applied Physics Letter*, Vol. 40, no. 40, 352–354.

13. Precision Machining Methods for Ceramics 313

Vora, H.; Orent, T. W.; and Stokes, R. J. (1982). Mechanochemical polishing of silicon nitride. *Journal of the American Ceramic Society*, Vol. 65, no. 8, C-140–C-141.

Wada, R. (1982). Ultra-precision diamond lathe. *Bulletin of the Japan Society of Precision Engineering*, 16–1, 8–11.

Warmbrod, W. M. (1972). *SME Technical Paper*, MR 72–613.

Watanabe, J. et al. (1972). Denshi buhin yō kesshō zairyō no chōseimitsu rappingu [Ultra-precision lapping of crystalline materials for electronic components]. *Journal of the Japan Society of Precision Engineering*, 3811, 753–757.

Watanabe, J. et al. (1981). *Annals of the CIRP*, 30, 91.

Watanabe, J. (1983). *Denshi buhin yō kesshō zairyō no chōseimitsu rappingu* [Ultra-precision lapping of crystalline materials for electronic components]. *Ceramics Japan* 18, 491–497.

Whitten, L. G. and Lewis, T. G. (1966). Machining and measurement to submicron tolerance. In *Proceedings of the Seventh International Machine Tool Design and Research Conference*.

Willimans, R. L. and Warmbrod, W. M. (1973). Mirror-like finishes achieved by microinch machining. *Cutting and Tool Engineering*, (Nov./Dec.).

Yasunaga, N. et al. (1975). *Kōseido*.

Yasunaga, N. (1980). *Seramikkusu no rappingu, porishingu kako* [Lapping and polishing of ceramics]. *Kgyō zairyō*, 28, 34–38.

Yasunaga, N. (1984). *Seramikkusu no rēza kako* [Laser machining of ceramics]. *Kikai gijutsu*, 32, 61–65.

Yasunaga, N. (1985). *Seramikkusu no rēza kakō gijutsu* [Laser machining techniques for ceramics]. *FC Repōto*, 3, 11–20.

Yoshinaga, H. et al. (1980). *Seramikkusu no setsudan kakō* [Cut-off machining of ceramics]. *Kōgyō zairyō*, 28, 23–28.

APPENDIX

Chronology of the Development of Advanced Ceramics

Suezo Sugaike
Sci-Tech Research Co., Ltd.
Chiyoda-ku, Tokyo 100, Japan

YEAR	ELECTRONIC CERAMICS	ENGINEERING CERAMICS	BASIC SCIENCE AND TECHNOLOGY IN GENERAL
1893		E. G. Acheson patented his synthetic SiC (U.S. Patent No. 492, 767, February 28, 1893). In an attempt to make a synthetic diamond, he heated clay and coke in the electric furnace. There he discovered SiC single crystals.	The diesel engine was invented (Germany, R. Diesel).
1903		Germany's Bosch made spark plugs principally composed of talc.	
1904		F. Boelling tried an admixture of boric acid and alkali feldspar in sintering SiC, receiving German patent 118,008 (May 22, 1904).	The diode vacuum tube was invented (Great, Britain, J. A. Fleming).

Note: The distinction between electronic and engineering ceramics is not applicable during the early years of the development of ceramic technology.

YEAR	ELECTRONIC CERAMICS	ENGINEERING CERAMICS	BASIC SCIENCE AND TECHNOLOGY IN GENERAL
1906		Japan's Shōfū Toki Ltd. was founded in Kyoto to manufacture insulators. The first insulators produced in Japan were made by Fukagawa Eizaemon VIII in 1871, as ordered by the Ministry of Public Works. Fukagawa founded Koran-sha (in Arita, Saga Prefecture) in 1879.	Theory of specific heat of solid bodies stated (Germany, A. Einstein).
1909		Japan's Asahi Glass made plate glass by machine blowing of glass cylinders. The firm was founded in 1907.	
1910		The German firm Prometheus made its SiC heating element (resistance element) commercially available under the name Silundum.	
1912			X-ray diffraction by crystals was confirmed (Germany, M. v. Laue). Bragg's equation also was published (Great Britain, W. H. Bragg).
1913		A patent was granted to the Thomson-Houston Co. for making tools, dies, rolls, and bearings of sintered alumina (Al_2O_3) (British patent 4887, February 27, 1913). Practical use of the substance did not begin until the 1950s, however.	

Appendix: Chronology   317

YEAR	ELECTRONIC CERAMICS	ENGINEERING CERAMICS	BASIC SCIENCE AND TECHNOLOGY IN GENERAL
1918		Japan-American Sheet Glass started up with technical contributions from the U. S. firm Libbey Owens. Sheet glass was produced, by drawing the glass at high temperatures from a vat of molten glass. The firm changed its name to Nippon Sheet Glass in 1931.	Wireless communication began to use vaccum tubes instead of spark transmitters.
1919		Nippon Toki (now Noritake) perfected a production technique for special high-voltage insulators. The firm's insulator division was made independent as NGK Insulators.	
1920	(The first radio broadcasts were begun in the United States by Westinghouse.)	Great Britain's A. A. Griffith recognized, from observation of the fracture of solids, the relationship between the concentration of stress on a crack and its development. Between 1920 and 1924 he developed his theory of the fracture of brittle materials, which is the basis for research into the strength of ceramics today.	The high level of strength of glass fibers was pointed out (Great Britain, A. A. Griffith).
1927	The high-frequency insulating qualities of steatite porcelain ($MgSiO_3$) were recognized (Germany).		The diffraction of an electron beam by a crystal was demonstrated, establishing the wave motion of electrons (United States, C. J. Davisson, L. H. Germer).

YEAR	ELECTRONIC CERAMICS	ENGINEERING CERAMICS	BASIC SCIENCE AND TECHNOLOGY IN GENERAL
1930	Cu–Zn ferrite and Co ferrite magnets were discovered (Tokyo Institute of Technology, Katō Yogorō and Takei Takeshi). (The worldwide depression had begun the previous year.)	NGK Insulators mass-produced spark plugs of mullite ($Al_6Si_2O_{13}$) porcelain. In 1936 that division of NGK Insulators became NGK Spark Plug.	
1931		In Germany spark plugs of alumina (Al_2O_3) ceramics were announced.	The theory of electrical conduction in semiconductors was published (Great Britain, A. H. Wilson).
1932		In Europe and the United States, SiC heating elements were effective in melting metals, with performance under actual use up to 1500°C. At present it is stable under use up to 1600°C.	
1933	The development of TiO_2 ceramic condensers occurred, principally in Germany.	Germany's Siemens marketed alumina porcelain vessels under the brand name Sinterkorund. O. Ruff had been developing the new product from 1924 on.	
1934	The Japanese government's Electrical Test Laboratory (renamed the Electrotechnical Laboratory in 1960), the University of Tokyo, the Tokyo Institute of Technology, and other institutions began research on steatite porcelains.	(The Japanese automobile industry began.)	

Appendix: Chronology 319

YEAR	ELECTRONIC CERAMICS	ENGINEERING CERAMICS	BASIC SCIENCE AND TECHNOLOGY IN GENERAL
1935	Initial research on TiO_2 ceramics began at the Electrical Test Laboratory in Japan. $CaTiO_3$ ceramics were also studied there. TDK (Tokyo Denki Kagaku Kogyo) was founded to produce soft ferrite.	I. W. Kurtschatow et al. of the Soviet Union discovered remarkable voltage-dependent nonlinearity with electrical resistance in a sintering body of SiC powder and clay. This discovery is the forerunner of today's varistor and arrester (electronic ceramic-related) applications.	
1936	At the German firm Osram, TiO_2 50-MgO 50 porcelains and kaolin porcelains were used as vaccum tube bases, with the glass portion sealed. Anticipating growing demand for TiO_2 as a raw material, Titan Kogyo Co. was established in Japan.		A cyclotron was built at the Institute for Physical and Chemical Research (Nishina Yoshio and colleagues). Nishina was the leader of the development project. His achievements are described in *Kagaku* (1956).
1937	(At this period, the production of radios in Japan had reached 400,000 a year.)	Trial production of SiC began in Japan. Tokyo Denki (now Toshiba) used it as a heat source for melting aluminium alloys for aircraft use for military demand. Tokai Denkyoku Seizo and Toho Sangyo Kenkyujo also started research to meet military demand.	(In North China, near the Chinese Bridge, the war between China and Japan started.)
1938		A. J. Thompsom of the U.S. firm Carborundum commenced production of SiC heating elements by the reaction sintering (recrystallization) method at 2300° to 2500°C (U.S. Patent No. 2,188,693, January 30, 1940).	Perfection of the commercial electron microscope (Germany, E. Ruska and B. von Borries of Siemens).

YEAR	ELECTRONIC CERAMICS	ENGINEERING CERAMICS	BASIC SCIENCE AND TECHNOLOGY IN GENERAL
1939	Production of TiO_2 ceramic condensors began in Japan at the Kawabata Seisakusho and elsewhere.	(World War II begins.)	Basic theory of the fracture rate of brittle materials stated (Switzerland, W. Weibull).
1941	With steatite ceramics used widely in vaccum tubes and insulators for radios, the industry expanded. Manufacturers included Kobe Kogyo, Kawabata Seisakusho, Toshiba, NGK Insulators, NEC, Mikasa Denki, Daito Seisakusho, Aoi Musen, and Shueki Kogyo. Production of TiO_2 ceramic condensors also expanded. Kawabata Seisakusho, Toshiba, and NEC all produced them. Ferrite was used in components for naval radio equipment. Commercial production of it by what is now the TDK Corporation was successful.	Field trials of spark plugs for airplanes made of high-density alumina began at Hitachi, Ltd. As a result, production began the following year (1942). Other manufacturers joining in their production included NGK Spark Plug and Aichi Kagaku Kogyo. (Japanese domestic automobile production reached 42,800 vehicles a year. In 1936, when the government took action to promote production—particularly of army vehicles—production was somewhat more than 1,000 vehicles a year.) (In December, the United States entered World War II against the Axis nations.)	In England, rader became practicable. In the previous year, a patent was granted to Nagai Kenzō of Tohoku University for his discovery of the alternating current magnetic biasing method, now widely used in magnetic recording heads.
1942	Research began into the synthesis of mica (Nagoya University).	In Japan, the favorable strength at high temperatures and great thermal conductivity of SiC refractories had been recognized, but their high cost meant they were mainly used only in laboratory applications. Specific gravity 2.1 (degree of sintering 65%).	E. Fermi achieved a controlled, sustained nuclear chain reaction.

Appendix: Chronology   321

YEAR	ELECTRONIC CERAMICS	ENGINEERING CERAMICS	BASIC SCIENCE AND TECHNOLOGY IN GENERAL
1943	Development began of ferrite magnetostrictive oscillators and electromagnetic radiation absorbers; research on magnetic recording tape started.	In Germany, alumina envelope vaccum tubes were being manufactured (by Telefunken). Alumina ceramics had developed to withstand usage in walls under a vaccum.	The ferroelectricity of $BaTiO_3$ had been discovered in a few countries (in the United States by E. Wainer, and A. N. Salomon; in the USSR, by B. W. Wul and I. M. Goldman).
1944	The ferroelectricity of $BaTiO_3$ was discovered in Japan at the Denki Shikenjo (electric test laboratory) by Ogawa Takeo.	Nippon Denko (now Toshiba Ceramics) began production of SiC heating elements.	
1945		Tokai Konetsu Kogyo began production of SiC heating elements, taking over from Tokai Denkyoku Seizo KK.	(World War II ended.)
1946	In Germany, G. Busch and H. Labhart studied the semiconducting characteristics of single-crystal SiC. Coarse crystals were used as experimental material, in the Acheson method.		
1947	Demand expanded greatly for ceramic condensers and ferrite magnetic cores. The U.S. Army of Occupation in Japan ordered production of superheterodyne radios. The piezoelectric effects in $BaTiO_3$ ceramics was verified (United States, S. Roberts).	Chubu Sangyo (which became Siliconitto Konetsu Kogyo in 1951) began working on commercial production of the SiC heating elements developed by Toho Sangyo Kenkyujo (dissolved after World War II).	The conception of electron valence control in oxide semiconductors was published (E. J. Verwey, The Netherlands).

YEAR	ELECTRONIC CERAMICS	ENGINEERING CERAMICS	BASIC SCIENCE AND TECHNOLOGY IN GENERAL
1948	The Néel theory of oxide ferrites was published (L. Néel, France). (Radio production had recovered to the prewar level in Japan.)	The possibility of making turbine vanes targeted for 1200°C using alumina ceramics or cermet in which alumina is the main component was explored (R. F. Geller, United States). This type of exploration continued throughout the 1950s.	The Ge transistor was discovered (W. Schockley, J. Bardeen, and W. H. Brattain, United States).
1950	The existence of antiferroelectricity in $PbZrO_3$ was pointed out (Takagi Yutaka, University of Tokyo). The idiosyncratic peak in the temperature-related changes in permittivity was found by measurement of the permittivity at the Electric Test Laboratory. Tests of baking circuits onto a thick film on a ceramic substrate appeared (Centralab and others, United States). The ceramic condenser manufacturers Murata Manufacturing and Taiyo Yuden were founded. TDK also moved into the ceramic condenser field.	The technique of sealing together ceramics and metals through baking with Mo–Mn appeared, with potential for applications not only in vaccum tubes but also in vaccum distributors. A method which adds Ti hydride to the joint metal as flux also was developed. It was reported that bending strength of about 206 MPa was achieved with a complex hot-pressed body of Al_2O_3-TiO_2 (H. N. Bar, G. D. Cremer, W. J. Koshuba). (With the promulgation of the Introduction of Foreign Capital law in Japan, the introduction of foreign technology into Japan became widespread.)	
1951	Research into ferrite memory cores was conducted in all directions.	(Toray Industries introduced technology for nylon from Du Pont.)	Work began on rebuilding the 4 MeV cyclotron at the Institute for Physical and Chemical Research in Japan.

Appendix: Chronology 323

YEAR	ELECTRONIC CERAMICS	ENGINEERING CERAMICS	BASIC SCIENCE AND TECHNOLOGY IN GENERAL
1951 (cont'd.)	The Langevin piezoelectric element was developed using $BaTiO_3$ ceramics and applied to fish detectors (Murata Manufacturing and Kyoto University cooperated.) $BaTiO_3$ ceramic pickups reached the commercial product stage (Rion Co.) The grain-boundary varistor effect was discovered in $ZnO-SnO_2$ ceramics (Toshiba).	(Commercial radio broadcasting began.)	
1952	Microwave applications using the Faraday effect and ferrite ferromagnetic resonance began. Hexagonal ferrite was reported. The types of magnetic oxides known were becoming increasingly varied (Philips, The Netherlands). Mn–Zn ferrite patented (Okamura Toshihiko, Tohoku University).		With the discovery of transistors, today's integrated circuits were anticipated by G. W. A. Dummer of Great Britain, who suggested a structure in which the electronic circuits form layers with resistors, conductors, detectors, and amplifiers combined, with the component areas separated as needed.
1953	A comprehensive phase diagram for $PbZrO_3$-$PbTiO_3$ was published (Sawaguchi Etsuro, Tokyo Institute of Technology). That later became basic data for the discovery of piezoelectric $Pb(Zr,Ti)O_3$ ceramics (PZT).		

YEAR	ELECTRONIC CERAMICS	ENGINEERING CERAMICS	BASIC SCIENCE AND TECHNOLOGY IN GENERAL
1953 (cont'd.)	NHK (Japan Broadcasting Corporation) and the commercial broadcaster Japan Television began television broadcasting in Japan. As television became widely popular, it developed into a new market for ceramic electronic components.		
1954	The PTC resistor was discovered by means of semiconducting $BaTiO_3$ (P. W. Haayman, et al., The Netherlands).		The discovery of masers (C. H. Townes, United States) opened up applications in microwave space communications. Al_2O_3/Cr^{3+} rubies were used.
1955	To develop semiconductors to follow the germanium and silicon devices, research on the production of highly pure single-crystal SiC was carried out (J. A. Lely, Germany).	Tests of the reaction sintering of Si_3N_4 were conducted, and its potential as a high-temperature refractory material was pointed out. The extremely small changes in volume that occur when objects are formed of Si powder and then sintered in a nitriding reaction were emphasized (J. F. Collins and R. W. Gerby, United States). That information later attracted great interest in Great Britain. A theoretical formulation for sintering contraction was based on the premise of volume diffusion. It shrinks at 2/5 of the heating time (W. D. Kingery and M. Berg, United States).	General Electric (United States) succeeded in creating synthetic diamonds (F. P. Bundy, H. M. Strong, and R. H. Wentorf, Jr.).

Appendix: Chronology 325

YEAR	ELECTRONIC CERAMICS	ENGINEERING CERAMICS	BASIC SCIENCE AND TECHNOLOGY IN GENERAL
1956	Garnet-type ferrites $Y_3Fe_5O_{12}$ (YIG) were reported (United States, France) (Transistor radios marketed by Sony.)	Experimental research into the porosity and bending strength of sintered alumina (R. L. Coble and W. D. Kingery, United States). Research into stabilizing ZrO_2 (Yamauchi Toshiyoshi and Shigeyuki Sōmiya, Tokyo Institute of Technology; Uei Isao, Nakazawa Yasuro, and Uetsuki Tooru, Kyoto Institute of Technology, Japan).	The Corder Hall nuclear power plant went on line (Great Britain).
1957	The U.S. firm Clevite's $Pb(Zr,Ti)O_3$ piezoelectric ceramic PZT replaced the earlier $BaTiO_3$ piezoelectric ceramic. High-frequency insulating uses were found for forsterite Mg_2SiO_4 ceramics.	In Germany, Patent No. 963, 766 was granted to an alumina cermet combination for use in turbine blades. It was 56% Al_2O_3, 4% Cr_2O_3, 30% Cr, and 10% Mo, sintered at 1700°C with a molding pressure of 28,000 psi (S. Tacvonan and M. Levecque).	It was demonstrated that stabilized ZrO_2 solid solution is a good oxygen ion electrolyte (K. Kiukkla and C. Wagner, Germany). (The Ezaki diode was invented by Ezaki Reona, Japan.) (The USSR put the world's first artificial satellite in orbit. The United States followed suit the following year.)
1958	The development of ferrite magnetic heads with abrasion resistance superior to that of metallic permaloy occurred. The excellent electrical insulating qualities of high-alumina content ceramics were pointed out (F. E. V. Spencer, D. Turner).	AlN brick was fired and its thermal and corrosion-resistant qualities recognized (M. Rey, Great Britain). The existence of α and β forms of Si_3N_4 was pointed out through Xray crystallography (E. T. Turkdogen et al., Great Britain).	The laser effect was predicted as a development of the maser effect with visible light (A. L. Schawlow and C. H. Townes, United States).

YEAR	ELECTRONIC CERAMICS	ENGINEERING CERAMICS	BASIC SCIENCE AND TECHNOLOGY IN GENERAL
1959	Basic research was conducted on the PTC characteristics of $BaTiO_3$ semiconductors (Saburi Osamu, Japan). In the United States, particularly at Bell Laboratories, research began on growing single crystals expected to be used in lasers ($Y_3Al_5O_{12}$, $CaWO_4$) or to have optical modulation and amplification uses (the BaTiO group of complex oxides) (P. C. Linares, J. W. Nielsen, J. P. Remeika, et al.). Kyoto Ceramics (Kyocera) was founded.	Success in developing translucent alumina ceramics (Lucalox) (General Electric, United States). The British Admirality Materials Laboratory published a comprehensive report (No. A175) of the development of Si_3N_4 (N. L. Parr et al.). U.K. Defence Laboratories perceived its potential as a material for gas turbine parts. It was pointed that β-Si_3N_4, like Ge_3N_4, is a phenacite (Be_2SiO_4)-type crystal. Its affinity to nitrides and silicate minerals became clear (D. Hardie and K. H. Jack, Great Britain). The analysis of the ceramics sintering phenomenon with the existence of a liquid phase was published. (W. D. Kingery, United States.)	The conception of the technology for today's integrated circuits was taking shape, and prototypes began appearing (J. S. Kilby and B. Noyce, United States). At the same time, a shift from Ge to Si for transistor materials was also occurring.
1960	A patented barrier-layer capacitor was announced (Glove Union, Erie Resistor, National Research and Development, United States.) Production of magnetic recording tape in Japan took off.	AlN was hot-press sintered and its characteristics evaluated (K. M. Tayler, C. Lenie, Carborundum, United States). Use of hot-press techniques spread rapidly from this point on. (Color TV broadcasting began in Japan.)	Amorphous metals were first produced by the splat cooling method (P. Duwetz, United States). Pulse laser emission with a ruby (Al_2O_3/Cr^{3+}) single crystal was achieved (T. H. Maiman, United States).

Appendix: Chronology 327

YEAR	ELECTRONIC CERAMICS	ENGINEERING CERAMICS	BASIC SCIENCE AND TECHNOLOGY IN GENERAL
1961	Microelectronics technology flourished, aiming at miniaturization of electronic component circuitry. A theoretical explanation was given for the PTC special effect in $BaTiO_3$ semiconductors (W. Heywang, Germany).	The U.S. firm, Corning, succeeded in producing transparent crystallized glass, Pyroceram (S. D. Stookey). Later, interest focused on their work as one method of producing microcrystalline ferroelectrics and transparent substances for opto-electronics applications. G. G. Deeley et al. of Great Britain's Plessy discovered that MgO is an effective sintering additive in the hot-press sintereing of Si_3N_4. The sintering phenomena in the middle and final phases of the process of sintering ceramic powders were explained with a diffusion model (R. L. Coble, United States.)	
1962	A 455-kHz intermediate filter for radios based on a PZT piezoelectric ceramic was produced commercially. The trend to greater performance and nonregulation in circuitry was becoming practical. The effectiveness in obtaining perfect single crystals of oxides and compound oxides by the Czochralsky method, used as a production method for single crystals of Ge and Si, was pointed out. Tests of		Using $CaWO_4/Nd^{3+}$, normal-temperature continuous-wave lasers became possible (L. F. Johnson, et al., United States). General Electric and IBM (United States) discovered the gallium arsenide semiconductor laser. The field-effect transistor using Si became a reality; that became the foundation for today's MOS transistors (S. R. Hofstein and F. P. Heiman, RCA, United States).

YEAR	ELECTRONIC CERAMICS	ENGINEERING CERAMICS	BASIC SCIENCE AND TECHNOLOGY IN GENERAL
1962 (cont'd.)	single-crystal growth for $CaWO_4$, Al_2O_3, and YVO_4 were made (K. Nassau, L. G. van Uitart, and A. M. Broyer, Bell Laboratories, United States).		
1963	Production of PTC thermistors of $BaTiO_3$ semiconductor began (Murata Manufacturing and others, Japan). A patent application was submitted for a delay device using the surface waves of piezoelectric substances (J. H. Rowen, United States, U.S. Patent No. 3,289,144, granted November 29, 1966).		
1964	The use of glass fibers as the waveguide path for optical communications was considered (Standard Telecommunication Laboratory, Great Britain). The proposal was made to add graded index of refraction changes to the glass fibers to be used in optical communications to give them focusing capabilities (Nishizawa Jun'ichi and Sasaki Ichiuemon, Tohoku University, Japan). Microcrystals of $BaTiO_3$ (50 nm or smaller) were made by crystallization from glass (A. Herczog, United States).		

YEAR	ELECTRONIC CERAMICS	ENGINEERING CERAMICS	BASIC SCIENCE AND TECHNOLOGY IN GENERAL
1965	Production began of piezoelectric elements to ignite gas (NGK Spark Plug, Matsushita Electric Industrial, Lyon, etc., Japan). $Pb(Mg_{1/3}Nb_{2/3})O_3$-$Pb(Zr,Ti)O_3$ group piezoelectric ceramics were reported (Matsushita Electric Industrial). That stimulated the development of piezoelectric ceramics with many constituents. Production began of hot-pressed ferrite for ferrite magnetic heads (Matsushita Electric Industrial and others). Surface wave device experiments with comb electrodes opposite on a piezoelectric crystal were conducted (R. M. White and F. W. Voltmer, United States).	With the expectation that alumina ceramics would be used for artificial bones, joints, dental roots, etc., tests began with animals in Japan. The effect of bonding materials, additives, and the sintering atmosphere upon the sinterability and flexural strength of Si_3N_4 became a subject of research in Japan (Saitō Shinroku, Sōmiya Shigeyuki, and Katō Koichi, Tokyo Institute of Technology, Japan).	
1966	The technical advantages of using quartz glass in fiber optical communications were suggested (K. C. Kao and G. A. Hockham, Great Britain) Barrier-layer-type $BaTiO_3$ capacitors were developed (NTT Telecommunications Laboratory, Japan). A multimode filter forming a bandpass filter was developed making use of the two or more types of oscillations modes that exist in a single resonator (Onue Morio, University of Tokyo, Japan).	(Oil became the most important energy source. A series of petrochemical complexes appeared. Japan's population broke through the 100,000,000 mark.)	

YEAR	ELECTRONIC CERAMICS	ENGINEERING CERAMICS	BASIC SCIENCE AND TECHNOLOGY IN GENERAL
1967	Laminated ceramic condensors (chip capacitor) became the object of development, in response to circuit miniaturization accompanying development of the integrated circuits. β-alumina ceramics' high Na ion conduction was discovered, and the first experimental Na-S batteries were made (Ford Motor Company, United States).	The U.S. firm Tyco succeeded in the Czochralski pulling of a filament of sapphire single crystal, with a minimum diameter on the order of 0.10 mm (H. E. LaBell, Jr., A. I. Mlavsky). A six-year research program concerning AlN was put into action in Japan at the Science and Technology Agency's National Institute for Research in Inorganic Materials.	Bubble domains were successfully formed and their motion controlled. These have potential as a new memory device (A. H. Bobeck, United States).
1968	Varistor devices using ZnO ceramics were developed (Matsushita Electric Industrial). Demand for them arose for protecting transistor circuits, and export of the technology for them to the U.S. firm General Electric began in late 1970. Selfoc, a glass optical conductor with focusing power developed with ion exchange technology, was announced (joint development by Nippon Sheet Glass and NEC, Japan). Research was moving forward on ceramic dielectrics with low loss in the microwave region.		The U.S. firm RCA announced its development of a liquid-crystal display device.

YEAR	ELECTRONIC CERAMICS	ENGINEERING CERAMICS	BASIC SCIENCE AND TECHNOLOGY IN GENERAL
1969	In Japan and the United States, production began of ceramic multilayer packages, that is, containers for integrated circuits in which the metallic conductive layers are sandwiched between ceramic substrates and all fired at once (Kyoto Ceramics (Kyocera), NGK Spark Plug, and other firms). The ferroelectric, ferroelastic substance $Gd_2(MoO_4)_3$ was reported (Hitachi Ltd., Japan). The applications of the optoelectric effect of the translucent ceramic PLZT (PZT with La) were pointed out (C. E. Land and G. H. Haertling, United States). Doughnut-shaped Ni–Zn ferrite with an external diameter of 20 cm was exported to the United States for the National Accelerator Laboratory, for use in the proton accelerator (Toshiba).	With the sputtering of silicides of transition metals ($MoSi_2$, $TaSi_2$, etc.) on silicon integrated circuits forming Schottky barrier diodes, ohmic electrodes, and direct connections, the trend to high-speed circuits began to appear (Bell Laboratories, United States). For sputter sources, high-purity sintered bodies of these silicides are regarded as essential.	Apollo 11 landed on the moon; Commander Armstrong stood on the moon's surface (United States). Quartz oscillator wristwatches were commercially produced (Seikosha, Japan).
1970	Development of a surface wave filter with piezoelectric media was begun by several firms in Japan. A smoke detector appeared that exploited the increase in electrical conductivity due to gas absorption by the surface of a highly	Advanced Materials Engineering, Ltd. (AME) was established by the pooling of funds and technology from Great Britain's governmental bodies, ceramics firms, and automobile manufacturers for the production and sale of Si_3N_4 sintered bodies.	The continuous operation, at room temperature, of a short wavelength (0.8–0.9 μm) semiconductor laser became a reality (Bell Laboratories, United States). With it, the light source for short-distance optical transmissions was set.

YEAR	ELECTRONIC CERAMICS	ENGINEERING CERAMICS	BASIC SCIENCE AND TECHNOLOGY IN GENERAL
1970 (cont'd.)	porous SnO_2 semiconductor (Figaro Giken and others, Japan). Production technology for β-alumina solid electrolytic ceramics was moving ahead. Test production of Na-S batteries was reported (jointly developed by Toshiba and Yuasa Denchi). Corning Glass achieved a low loss (20 dB/km) optical communication fiber by chemical vapor deposition. That stimulated development of quartz glass fiber in many countries.	That made possible the supply of raw material powders of high purity for the development of gas turbine parts and heat transfer devices.	The semiconductor material was $(Al,Ga)As$. The double roll production method for amorphous metals was developed. The circuits for color TVs were modularized by type of circuit in an attempt to save labor in maintenance (RCA and other U.S. firms).
1971	Continuous pulling of alumina single crystals from a molten layer of alumina welling up to the surface of a capillary float, EFT (Edge-defined Film-fed Crystal Growth), was reported (H. E. LaBell, Jr. and A. I. Mlavsky, Tyco, United States). Depending on the state of the float, flat or hollow single crystals in a free range of shapes can be produced. Later this method was used industrially to produce sapphire sheets and silicon sheets for solar batteries. Barrier-layer-type $BaTiO_3$ nonlinear capacitors were announced (NTT's	Toshiba succeeded in producing high-density, high-strength, hot-pressed sintered bodies by adding Y_2O_3 to AlN (Komeya Katsutoshi and Inoue Hiroshi). That was the starting point for the use of Y_2O_3 as a sintering additive for all types of nitrides. In the United States, the Department of Defense's ARPA (Advanced Research Program Agency) commissioned research into ceramic materials for turbines and the development of ceramic turbine parts. The research centered on 30 MW deferred-type turbines (Westinghouse) and	

Appendix: Chronology   333

YEAR	ELECTRONIC CERAMICS	ENGINEERING CERAMICS	BASIC SCIENCE AND TECHNOLOGY IN GENERAL
1971 (cont'd.)	Musashino and Ibaraki telecommunications laboratories, Japan). Their capacity changes nonlinearly according to the direct current bias voltage. A switch was being made from ferrite core memory to semiconductor integraded circuit memory for use in computer main memory.	automobile turbines (Ford Motor Company).	
1972	Production by Japanese manufacturers of PTC thermistors of $BaTiO_3$ semiconductor expanded. Main applications include temperature control in home appliances, such as the automatic rice cooker, and demagnetization current control in color TVs.	It was experimentally shown that the thermal shock resistance of partially stabilized ZrO_2 ceramics (PSZ) (measured by C. E. Curtis in 1947) is due to the mixing in of microscopic monoclinic crystals among the isotropic ZrO_2 matrix grains (R. C. Garvie and P. S. Nicholson, Australia). In Japan, experiments with animals, leading to applications of alumina single crystals (sapphire) as artificial dental roots and artificial bone material began (Kyoto Ceramics (Kyocera), Osaka Dental College). In the United States, NASA began experiments in using Si_3N_4 spheres as bearings.	

YEAR	ELECTRONIC CERAMICS	ENGINEERING CERAMICS	BASIC SCIENCE AND TECHNOLOGY IN GENERAL
1973	Oxygen sensors of stabilized ZrO_2 solid electrolyte were demonstrated to indicate accurately the air–fuel ratio in an automobile engine (R. Zechnall, G. Baumann, and H. Fisele, United States.) A transparent Y_2O_3 ceramic with Nd_2O_3 in solid solution was produced and made to function as a laser (C. Greskovich and J. P. Chernoch of General Electric, United States).	It was reported that the strength at high temperatures of Si_3N_4 hot-pressed sintered bodies can be improved by the addition of Y_2O_3 (G. E. Gazza, United States). The explanation was that an yttrium silicate glass grain-boundary layer is formed that is superior in heat resistance to the conventional MgO enstatite glass grain-boundary layer.	
1974	Bell Laboratories (United States) reported achieving quartz glass fibers for optical communications with the low loss level of 5dB/km with their modified CVD method (MCVD) (J. B. MacChesney et al., Bell Laboratories, United States). The technology was developed to make alumina substrates of high purity and with the planar smoothness of glass (surface roughness of 0.05 μm or less, and 70–80% of the glossiness of glass). They are useful as substrates for hybrid integrated circuits (Fujitsu Laboratories, Ltd., Japan). A $BaTiO_3$ ceramic capacitor, with high permittivity in the high-frequency region and that can withstand high voltages, was	Toshiba succeeded in crystalizing the grain-boundary phase of Si_3N_4 sintered body by using a Y_2O_3 and Al_2O_3 additive and hot pressing it. The result was markedly greater strength at high temperatures (Komeya Katsutoshi and Kudō Haruo, Tsuge Akihiko, Japan). Toyota laboratories pointed out that a sintered body of Si_3N_4 with AlN and Al_2O_3 in solid solution is advantageous in terms of resistance to oxidation, abrasion resistance, and low expansion. At about the same time, research into Sialon (Si-Al-O-N group nitrous oxide substances) became widespread internationally (Oyama Yoichi, Japan). Great Britain's British Nuclear Fuel Ltd.	(Japan suffered real negative economic growth, due to the oil scarcity caused by OPEC's 1973 decision to restrict petroleum output.)

Appendix: Chronology 335

YEAR	ELECTRONIC CERAMICS	ENGINEERING CERAMICS	BASIC SCIENCE AND TECHNOLOGY IN GENERAL
1974 (cont'd.)	developed (NTT's Ibaraki Telecommunications Laboratory, Japan). Permittivity is 3,000–4,000 at room temperature, changes due to temperature range from −1 to +2%. With rare earth (Dy,Gd) oxides as additives, 1 μm grains were achieved.	(BNFL) developed and began marketing reaction-sintered SiC ceramics (Refel). General Electric (United States) succeeded in pressureless sintering of high-density, high-strength SiC by using a refined β-SiC fine powder from which oxygen content had been excluded, adding boron, and controlling the free carbon content (S. Prochazka). The West German firm Annawerk reported development of ball bearings, roller bearings, and races of hot-pressed Si_3N_4 (E. Gugel et al.) A West German joint public and private national project to develop ceramic parts for gas turbines got underway (Prof. W. G. J. Bunk, general director).	
1975	Numerical displays using transparant PLZT ceramics were developed (Sandia Laboratories, United States). Long service life, response from 10 to 50 μs, and low power demand for holding memory are potentially valuable characteristics. Development of microwave dielectrics of $Ba_2Ti_9O_{20}$, Li_2O-TiO_2-Al_2O_3 group ceramics occurred. (Production of VTRs began to rise	At Tohoku University's Kinzoku Zairyo Kenkyujo (Iron, Steel, and Other Metals Institute), spun polycarbosilane was heated to 800 to 1300°C in a vacuum to produce continuous filament from a fine crystalline β-SiC aggregate (Yajima Seishi, Japan). Its tensile strength reaches 350 kg/mm². The phenomenon of the remarkable toughness of partially stabilized ZrO_2 (PSZ) was explained as due to	

YEAR	ELECTRONIC CERAMICS	ENGINEERING CERAMICS	BASIC SCIENCE AND TECHNOLOGY IN GENERAL
1975 (cont'd.)	gradually. In 1981, the value of VTRs produced in Japan surpassed that of color TVs to top one trillion yen.)	the absorption of stress energy by the tetragonal system fine grains (<30 nm) that exist metastably in it so that the ZrO_2 undergoes as martensitic transformation into its monoclinic phase (R. C. Garvie, R. H. Hannink, and R. T. Pascoe, Australia).	
1976	The development of circuit elements using thin-film technologies such as high-frequency sputtering became widespread. Accompanying that trend, demand for high-purity vaporization source oxide ceramics appeared. W. Heywang's explanation of the PTC characteristics of $BaTiO_3$ was further quantified and the great diffusion of empty oxygen lattice points in the $BaTiO_3$ grains was pointed out (J. Daniels et al., The Netherlands).	The U.S. Department of Energy's CATE (Ceramic Application for Turbine Engine) project got underway. General Motors' DDA division played a leading role, indicating applications in buses and trucks. The U.S. firm Carborundum announced it had succeeded in producing α-SiC powder which made possible pressureless sintered bodies of α with no loss of strength at 1650°C (J. A. Coppola and C. H. McMurty). In late 1977, the firm decided to establish research facilities to make this development commercially viable. A simple technique was suggested for rapid evaluasion of the stress intensity factor on small samples, using analysis of the indentation fracture in ceramic materials (A. G. Evans et al., United States).	With the success of the normal-temperature, continuous GaAs semiconductor laser using 1.0 to 1.7 μm wavelength, the way to long-distance, large-volume optical telecommunications was opened up (MIT, United States).

YEAR	ELECTRONIC CERAMICS	ENGINEERING CERAMICS	BASIC SCIENCE AND TECHNOLOGY IN GENERAL
1977	Amorphous $LiNbO_3$ was made by rapid cooling, and its unusually high permittivity and peak due to temperature changes were discovered. Then in 1978 it was pointed out that the Li ion electrical conductivity becomes a surprisingly large 10^{-3} $(\Omega \cdot cm)^{-1}$ (A. M. Glass et al., Bell Laboratories, United States). The same methods as used for amorphous metals extended to ceramic materials. In Japan, NTT's Ibaraki telecommunications laboratory led in reporting success with VAD, a new method of producing quartz glass fibers for optical communications (Izawa Tatsuo, Miyashita Tadashi, and Hanawa Fumiaki). Production of surface wave filter devices using $LiTaO_3$ single-crystal sheets began (Toshiba, Japan). The new device replaced more than ten component circuits conventionally used as a miniature intermediate frequency filter in TV. Surface wave filters of PZT group ceramic piezoelectric sheets were also commercially produced. They were used in 10.7 MHz FM	The U.S. Air Force commissioned Garrett to conduct research on the technology for making the vanes on the rotors and stators of turbine engines out of ceramics. The U.S. firms G.T.E. Sylvania began marketing Si_3N_4 powder suitable for making sintered bodies with good performance.	

YEAR	ELECTRONIC CERAMICS	ENGINEERING CERAMICS	BASIC SCIENCE AND TECHNOLOGY IN GENERAL
1977 (cont'd.)	tuners, 58 MHz TVs, and other applications (Murata Manufacturing, Mitsumi Electric, and other firms, Japan). Chip-type laminated capacitors began to be used as miniaturized components on circuit substrates along with silicon integrated circuits.		
1978	The ZnO varistors widely used in telecommunications devices and household appliances made inroads as electric power lightning arresters. Development extended to those for 500 kV electric power transmission circuits (Meidensha Electric, Mitsubishi Electric, and other firms in Japan). Higher pressure sodium vapor lamps using translucent alumina as their outer envelope began to be widely used in factory and street lighting as an energy-saving light source. Doorbells using piezoelectric ceramics were produced. In Japan, NTT conducted on-site tests of short-distance fiber optical transmission using cable of quartz glass fiber domestically produced by the MCVD method. The cable was strung over	In Japan, MITI's major system to develop energy-conservation technology, the Moonlight Project, took up high-efficiency gas turbine technology in its seven-year plan. The plan pointed to the use of ceramic parts, and goals were set for ceramics with three-point bending strength of at least 100 kg/mm^2 at room temperature and 60 kg/mm^2 at 1500°C. In a related move, the High Efficiency Gas Turbine Technology Research Consortium was formed in Japan on October 1, 1978. In the United States, contracts for development of ceramic engine parts were let by the Department of Energy, ARPA, the Air Force, the Army, NASA, Navy, the NSF, and other governmental bodies to many firms, including Garrett, General Motors, Ford,	

Appendix: Chronology 339

YEAR	ELECTRONIC CERAMICS	ENGINEERING CERAMICS	BASIC SCIENCE AND TECHNOLOGY IN GENERAL
1978 (cont'd.)	21 km between the control telephone repeater stations at Karagasaki and Hamacho in Tokyo. The Japanese firm Hoya supplied Lawrence Livermore National Laboratory in the United States with large pieces of phosphorofluoridate glass for lasers it had repeatedly improved. Inferring from the infrared permeability of oxide glass, S. H. L. Goodman of Great Britain pointed out that halide and chalcogenide glass would be superior to quartz glass in the 2 to 10 μm wave band.	General Electric, International Harvester, Martin-Marietta, Norton, and SKF. In Japan, in response to a movement to increase production efficiency on metal processing production lines, particularly in the automobile industry, the use of ceramic tools with alumina as their principal constituent suddenly began to spread. The Swedish firm ASEA reported a method of isostatically hot-pressed sintering of Si_3N_4. Sintered at 1700°C and 2 K bar, the surface finish of the Si_3N_4 was good, eliminating the need for post-sintering processing.	
1979	In electronic components, the automation of printed circuit production was a key point in television and radio manufacture. In ceramic condensers (capacitors), the conventional disk (radial) form received competition from the tubular (axial) form, the same form as resistors. Component manufacturers were also starting up development of automatic printing devices.	In Japan, synthetic Sialon was developed at the Science and Technology Agency's National Institute for Research in Inorganic Materials. With reaction sintering, it has a strength of 45 kg/mm^2. A joint project between Toshiba and Toshiba Ceramics succeeded in the production of an Si_3N_4 powder which, when hot-press-sintered, has over 126 kg/mm^2 at 1200°C.	

YEAR	ELECTRONIC CERAMICS	ENGINEERING CERAMICS	BASIC SCIENCE AND TECHNOLOGY IN GENERAL
1979 (cont'd.)		The University of California and Los Alamos Scientific Laboratory together completed their toroidal plasma vessel, the ZT-40 Torus, made of alumina ceramics. The external diameter of the doughnut is 320 cm. The U.S. Department of Energy inaugurated its Advanced Gas Turbine project. The DDA division of General Motors and Garrett play the central roles in this attempt to turn to ceramics for essential parts.	
1980	In Japan, NTT's Ibaraki telecommunications laboratory, using the VAD method, reduced the water content of quartz glass optical fiber and achieved the favorable characteristic of 0.5 dB/km (at 1.2 to 1.7 μm wavelength) transmission loss. The use of throwaway-type oxygen sensors for iron- and steelmaking, using solid electrolytic ZrO_2 ceramics stabilized with CaO, increased rapidly, from 30,000 units in 1977 to 190,000 units in 1980. Substrates for integrated circuits with high thermal conductivity, of SiC to which BeO is added, were developed. Their thermal conductivity is an order of magnitude higher than that of alumina substrates,	In a joint Kyoto Ceramics (Kyocera) and Isuzu Motors experiment, a ceramic experimental model of a 510 cm^3, 8-HP, single-cylinder diesel engine was built and tested. They succeeded in achieving more than 320 hours of operation. The piston cap, cylinder liner, and subcombustion chamber were made of Si_3N_4. Japan's Komatsu and the United States Cummins Engine jointly made and tested a turbo compound engine using ceramic parts (supplied by Toshiba and others). Si_3N_4 was the principal ceramic used. Japan's Ube Industries developed a process for making Si_3N_4 raw material	Japan's annual production of automobiles broke through the 10,000,000 unit level to become the world's greatest. Japan also led in iron and steel production in the free world. Japan continued to be the world's top automobile manufacturer in 1981.

Appendix: Chronology   341

YEAR	ELECTRONIC CERAMICS	ENGINEERING CERAMICS	BASIC SCIENCE AND TECHNOLOGY IN GENERAL
1980 (cont'd.)	0.7 cal/cm·sec·°C; their resistivity is 2×10^{13} Ω; and their thermal expansion is close to that of Si (Hitachi, Japan).	powders by a liquid-phase reaction between silicon tetrachloride and ammonia and moved ahead to build a pilot plant. Japan's Kurosaki Refractories developed a compound ceramic heat-exchange device of Si_3N_4 and SiC. The raw materials are high-purity silicon (99.99% and above) and an organic high molecular silicon compound. Asahi Glass (Japan) succeeded in developing Si_3N_4 and SiC products through pressureless sintering.	
1981	Denki Kagaku Kogyu produced LaB_6 single crystals using a floating zone melting method in a high-pressure argon atmosphere, for thermionic emission cathodes to be used in electron beam exposure devices in the production of LSIs. The firm made practical the method developed by the Science and Technology Agency's National Institute for Research in Inorganic Materials. Toshiba and Toshiba Ceramics jointly undertook to pull semiconductor-grade silicon single crystals with low oxygen content ($<2.10^{16}/cm^3$) from an Si_3N_4 crucible. The crucible shape was made of reaction-sintered Si_3N_4	The U.S. firm Garrett conducted cycle tests over 15 hours on ceramic rotor vanes for the T 76 turbine engine. The 25 metal vanes were replaced by Si_3N_4 vanes which successfully met two 7.5-hour trials. The turbine intake temperature was 1204°C. Japan's NGK Spark Plug began test operation of a 50 cm² two-cycle air-cooled engine in which all parts except the bearings and pistons were ceramic Si_3N_4 and Al_2O_3. It passed through a 100-hour no-load endurance test at 3,000 rpm. Japan's Toyo Soda Manufacturing began commercializing production of Si_3N_4 powders, using a	

YEAR	ELECTRONIC CERAMICS	ENGINEERING CERAMICS	BASIC SCIENCE AND TECHNOLOGY IN GENERAL
1981 (cont'd.)	or from pure carbon block. Its surface was coated with chemical vapor-deposited α-Si_3N_4 above 1300°C from $SiCl_4$ and NH_3 gas, fed in using N_2 gas as the carrier.	low-temperature liquid-phase surface reaction. Toray Industries of Japan achieved flexural strength of 150 to 170 kg/mm², superior to that of commercially available high-tension steel, with ZrO_2 ceramic partially stabilized with Y_2O_3 (PSZ). MITI's Research and Development Project of Basic Technologies for Future Industries got underway. Advanced ceramics was one of the areas of this R&D project. The goals of the ten-year project include developing high-strength materials, highly corrosion-resistant materials, and high-precision abrasion-resistant materials. To reach these goals, an advanced ceramics research consortium including 15 companies was formed (August 4, 1981). Kyoto Ceramics (Kyocera) and Fuji Kinzoku Kosaku jointly marketed all-ceramic pipe valves (with Al_2O_3 the principal ceramic used) to users in the corrosion-resistant, abrasion-resistant field.	
1982	Due to the development of the technology for spreading electrode paste on multiple sheets of ceramics and	On-site applications at iron and steel smelters and rolling mills, using Si_3N_4 and SiC as heat-resistant and	With the shift to continuous casting and thorough recovery of heat output, the energy required to

YEAR	ELECTRONIC CERAMICS	ENGINEERING CERAMICS	BASIC SCIENCE AND TECHNOLOGY IN GENERAL
1982 (cont'd.)	stacking them, a trend to laminated chip capacitors as well as laminated varistors, piezoelectric devicess, ceramic sensors, and heating elements appeared. The use of the hydrothermal process for the refining or manufacturing of primary particles of fine, high-purity oxides with favorable characteristics began to be considered. It is used with alumina, ferrite, stabilized zirconia, etc. (Sōmiya, Tokyo Institute of Technology, Research Laboratory of Engineering Materials, Japan, and Stambaugh, Battelle Memorial Institute, United States). Glass fibers using the infrared transmission of GeO_2 group glass were developed (Furukawa Electric, Japan). With loss of 13 dB/km at 2.05 μm, applications in radiant temperature measurement are being considered. Japan's KDD laid 50 km of optical fiber undersea cable in Sagami Bay and began experimental optical transmissions.	abrasion-resistant parts and as jigs, continued. Various bearing manufacturers in Japan also committed resources to the development of ceramic bearings. These included NSK, Fujikoshi, Koyo Seiko, NTN Toyo Bearing, Amatsuji Kokyu, and others. Development of an energy-saving diesel engine using ZrO_2 was carried out by Cummins Engine of the United States and NGK Insulators of Japan. A trend to private international exchanges in development appeared. An Si_3N_4 glow plug for a diesel engine was developed jointly by Isuzu Motors and Kyoto Ceramics (Kyocera). Nippon Tungsten, Japan, developed Si_3N_4 tools, pointing out their superior thermal shock resistance and omission resistance. Japan's Asahi Glass introduced the Refel SiC technology from Great Britain's BNFL (see 1974 for full name). General Electric in the United States instituted a silicon carbide products operation as a subsection of its industrial technology materials group.	produce a metric ton of steel was 85% of what was required in 1973.

REFERENCES

Dai-nippon Yōgyō Kyōkai (ed.). (1933). *Nihon yōgyō taikan* [Overview of the Janpanese ceramics industry]. Tokyo: Nihon Yōgyō Kyōkai.
Elmer, T. H. (1953). Silicon carbide heating elements—development, uses, chemical and physical investigations. *Bulletin of the Ceramic American Society* Vol. 32, no. 1, 23–25.
Gitzen, W. H. (ed.). (1970). *Alumina as a Ceramic Material.* Columbus, OH: The American Ceramic Society.
Kagaku gijutsushi nenpyō [Chronology of the history of science and technology] (1956). Tokyo: Heibonsha.
Kikan kōgyō rea metaru [Industrial rare metals quarterly] (1980). 73, 4–6.
Kirby, J. S. (1976). Invention of the integrated circuit. *IEEE Transactions on Electron Devices,* ED–23, 648–654.
Sōmiya, Shigeyuki (ed.). (1982). *Seramikkusu no kagaku to gijutsu no genjō to shōrai* [The present state and future prospects of ceramic science and technology]. Tokyo: Uchida Rokakuho Shinsha.
Sugaike, Suezo and Sōmiya, Shigeyuki. (1976). *Nihon ni okeru erekutoroniku seramikkusu no hattatsu* [Developments in electronic ceramics in Japan]. *Ceramics Japan,* 11, 638–645.
Yōgyō Kyōkai (ed.). (1951). *Nihon yōgyō taikan* [Overview of the Japanese ceramics industry]. Tokyo: Gihōdō.
Yōgyō Kyōkai (ed.). (1971). *Hatten suru seramikkusu* [Developing ceramics]. Tokyo: Yōgyō Kyōkai.

INDEX

A

Acoustooptic elements, glass for, 192–193
Active ceramics, definition, 8–9
ADP, growth, 204–205
Adsorption, chemical sensors, 168
Advanced ceramics
    definition, 8
    development chronology, 315–343
    processing, 20–21
Air-cooled turbine blades, 249, 252
Alumina, 31, 37–40, 120
    alkaline content and insulation resistance, 29–30
    artificial joints, 217–219
    beta, 120–121
    characteristics, 44
    cutting, 265–266
    dental and orthopedic applications, 217
    elastic emission machining method, 284
    fine-grained substrates, 107
    flowery-type, 41
    implants, 212–214
    laser drilling, 290
    manufacturing process, 31, 37–38
    mechanical properties, 216
    medical applications, 216–217
    particle size and impurities, 38–39
    polycrystalline, thermal conductivity, 157–158
    production factors, 38
    pulverized, 39
    sandy-type, 42
    types, 38–40
    ultrahigh-purity, 43
    uses, 45–46
Aluminum hydroxide, 39–40
Amperometry, 173, 176
Angle of diffraction, 192–193
Applied ceramics, structure, 4–5
ASTM-F603, 216

Automotive parts, 241–242
    future, 247–248
    honeycomb catalyst carriers, 244–245
    knock sensors, 246–247
    operating environment, 242
    oxygen sensors, 243–244
    riser heaters, 246
    thermo sensors, 245–246

B

Ball mill, 69
$Ba_2NaNb_5O_{15}$, 203–204
Barium ferrite, 134, 138–139
Barium titanate
    riser heaters, 246
    semiconductors
        current-time characteristics, 112
        temperature vs. resistance characteristics, 111
    varistor, 114
Basic ceramics, structure, 4–5
Batch kiln, 75
Batteries
    high-temperature use, 177–178
    room-temperature use, 178–179
Bayer process, 31, 37–38
Bending tests, 230–231
Beryllium oxide, 107
Binder, 70
Bioceram
    artificial bones and joints, applications, 220
    single-crystal alumina, 213–214
Bioceramics, 8
Biodegradable ceramics, 213
Bioinert oxide ceramics, 212–214
Biological applications, *see also* Medical applications
    metallic implants, 209–214

345

Body preparation, 66–69
  drying and granulating, 69
  mixing, 67–69
  powdered material selection, 67
Boron nitride, 59, 61–63
  insulating properties, 104
  production, 59–61
  properties, 61, 63
  structure, 61–62
  used, 63
Boundary-layer capacitors, 113–114
Breaking strength, 226–230
  thermal shock temperature differential and, 97–98
Brillouin zone, 152–153
Brittlenesss, 233

C

Calendar roll method, 15–16
Capacitor, 113
  boundary-layer, 113–114
Carbon
  crystalline and vitreous, 212
  medical applications, 214–216
  seals, engines, 253–254
Carboreduction, silica, 51
Catalysts, 183–184
Catalytic combustion-type gas sensors, 169
Centrifugal sedimentation method, 53
Chalcogenide glass, optical fibers, 197–199
Characteristics, ceramic, 28–29, *see also* specific properties
  commercially available aluminas, 44
  commercially available Ni_3N_4 fine powders, 59–60
  commercially available SiC fine powders, 52
  powders, 30–31
  substrate materials, 105–106
Chemical batteries, *see* Batteries
Chemical properties, 167–168
  applications, 167
  catalysts, 183–184
  chemical sensors, *see* Chemical sensors
  electrochromism, 180–182
  high-temperature steam electrolysis, 181, 183
Chemical pumps, 179–180

Chemical sensors, 168–176
  adsorption, 168
  amperometry, 173, 176
  approaches to selectivity, 168
  concentration cell, 170
  electromotive force, 170, 173
  gas sensor, 169
  humidity sensor, 168
  limiting current, 176
  oxygen sensor, 170–171
  potentiometric sensors, 172–174
Chemically bonded ceramics, definition, 9
Classification, 211–212
  by chemical composition, 11–12
  method of heating or sintering, 17–19
  by minerals, 12
  by molding technique, 12–16
  by properties, 23
CO_2 lasers, 288
  cutting, 293–294
  scribing, 295
Coal-gas fuel cell, 178
Coefficient of elasticity, 225
Concentration cell, 170
Conductor, three-dimensional structured, 121–122
Continuous kiln, 75, 78
Copiers, 197–198
Cordierite crystals, thermal expansion, 162–163
Cores, air-cooled turbines, 249, 252
Crack, growth, 97
Creep strength, 256
Cristobalite porcelain, 104
Critical temperatures, superconductors, 115–116
Crystal pulling method, 204
Crystalline carbon, implants, 212
Crystalline ceramics, definition, 6–7
Crystals, 199–200
  applications, 205–207
  CZ method, 204
  Helmholtz free energy, 201
  with light-emitting functions, 202–203
  light modulation, 205–206
  momentum, 152
  optical applications, 200
  optical properties, 199–202
  polarization, 201–202
  raw material melting, 204
  refractive index, 201
Curie point, 139–140

Index **347**

Cut-off methods, 302–307
  ceramics, 304
  other materials, 305–306
  single-crystal materials, 302–303
  using hydrostatic pressure, 303, 305, 307
Cutting, 264–265
  lasers, 291, 293–294
  pre-sintered ceramics, 264
  sintered ceramics, 265–266
  ultraprecision diamond cutting tools, 266–269
CZ method, 204

D

Debye's specific temperature, 148
Debye's theory of specific heat, 147
Definitions, 5–9
Deflection, 231
Deformation, 224–226
  strength, 226–230
Diamond
  cutting tools, 266–269
    advantages, 268–269
    dynamic balance, 268
    history, 270–271
    sharpness, 267
    temperature control, 268
    tool accuracy, 267
    vibration, 267–268
  dies, starting holes, 288, 290
  scribing, 293
  wheels, 273–277
    cast-iron bond, 275–276
    desired conditions, 273
    ground surface roughness, 276–277
    porous metal bond, 275
    standards, 275–276
    structure, 275
    system, 274
Die pressing, 12–13
Diffraction, 192–193
Display terminal, 181–182
Doctor blade method, 14–15, 72–74
Double cantilever method, 92–93
Double torsion method, 92–93
Drilling, laser, 288, 290–291
  efficiency increase, 288, 292
  machined volume, 291, 293
Drying, 69
Dulong-Petit's law, 147

E

Elastic constants, 89–91
Elastic emission machining method, 284
Electrical discharge machining, 307
Electrical properties, 36
  high-conductive ceramics, 114–115
  insulation, 104–109
  ionic conduction, 118–122
  semiconductors, 109–114
  superconductors, 115–118
Electric discharge complex grinding, 307–308
Electric double-layer capacitors, 176, 179
Electroceramics, 8
Electrochemical Knudsen cell, 180
Electrochromic displays, 181–182
Electrofused magnesia, 48
Electrolysis, high-temperature steam, 181, 183
Electrolytic complex grinding, 308
Electromotive force, 170, 173
Electronic ceramics, 8, 103
  development chronology, 315–343
Electrostatic bond strength, 163, 165
Engineering ceramics
  applications, 238
  definition, 8
  development chronology, 317–343
Engines, 248
  air-cooled turbine blades, 249, 252
  carbon seals, 253–254
  gas turbine, 248–251
  high-temperature materials, 255–256
  thermal barrier coatings, 253–254
  turbine shrouds, 255
Etching, laser-assisted, 297–300
Excimer laser, 299
Extrusion molding, 13–14, 74–75

F

Faraday rotation glass, 191–192
Ferrite, 126
  Al-substituted Mg-Mn, 139–140
  barium, 138–139
  circular polarization permeability, 129
  hard, 134–136
  history, 125
  magnetic characteristics, 127–128
  maximum energy product, 135

Ferrite (*continued*)
  microwave
    applications, 139–142
    characteristics, 128–129
  semihard magnetic, 136–139
  soft magnetic, *see* Soft magnetic ferrite
  static magnetic field, 129
  types, 126–127
Fine ceramics, definition, 7
Finishing, 80–81
Firing, 75, 78
  finishing, 80–81
  pressure sintering, 79–80
  reaction sintering, 78–79
Flame fusion method, 204
Flexural strength, 84–88
  grinding wheel grain size effect, 84–85
  high temperatures, 87–88
  shape and dimension effects, 86–87
  surface condition effect, 84–86
  test method, 84–85
Fracture mechanics tests, 231–232
Fracture strength, 225
Fracture toughness, 91–93, 232
Fuel cell, 119–120

G

GaAs substrates, 108
Gas-phase pyrolysis, 51
Gas sensor, 169
Gas turbine, 248–251
  carbon seals, 253–254
  development projects, 249–251
  thermal barrier coatings, 253–254
  turbine shrouds, 255
Glass, *see also* Optical fibers
  for acoustooptic elements, 192–193
  Faraday rotation, 191–192
  laser, 189–190
  laser amplification, 190–191
  photochromic, 192–194
  quartz group, 195
Glass membrane electrode, 172
Granulating, 69
Green body
  character, 28
  density changes, 55
Grinding, 269
  diamond wheels suitable for, 273–277
  electric discharge complex, 307–308
  electrolytic complex, 308
  energy, 273
  grain depth of cut, 272
  machine tools, 277–279
  mechanism, 269, 272–273
  model, 272
  precision rotary surface grinder, 278–279
  precision vertical surface grinder, 280
  ratio, 273
  surface grinder for hard, brittle material, 277–278
Ground surface roughness, 276–277
Grüneisen constant, 150
Gyromagnetic phenomenon, 128–129, 139

H

Hard ferrite, 134–136
Hardness, 90–91
Heat capacity, 94
Heating, classification by method, 17–19
Helmholtz free energy, 201
High-conductive ceramics, 114–115
High-performance ceramics, definition, 8–9
High-technology ceramics, definition, 8
High-temperature materials, 255–256
High-value-added ceramics, definition, 8–9
High-voltage applications, insulation materials, 104
Hip joint, replacement, 218
History, 3
Honeycomb catalyst carriers, 244–245
Hot-press sintering, 79–80
Humidity sensor, 168
Hydration machining, 284–285
Hydrothermal ceramics, 8

I

Implants
  biodegradable ceramics, 213
  bioinert oxide ceramics, 212–214
  carbon and ceramic systems, 211–214
  crystalline and vitreous carbons, 212
  metallic, 209–210
  optimizing, 211
  surface-active ceramics, 214
Impurities, effects, 29–30
Injection molding, 14, 71–72
Insulating properties, 104–109
  high-voltage applications, 104
  substrates, *see* Substrates

Integrated circuit packages, production, 73–74
Ion beam machining, 308
Ion concentration sensors, 172–173
Ionic bonding, 211
Ionic conduction
  beta-alumina, 120–121
  stabilized zirconia, 118–120
  three-dimensional structured conductors, 120–121
Ion selective electrodes, properties, 174–175
Isostatic hot-press sintering, 79
Isostatic pressing, 12–13, 69–71

J

JIS B 7726, 91
JIS R 1601, 84–85
JIS Z 2245, 91
JIS Z 2251, 91
Josephson tunnel junction devices, 116–117

K

KDP, growth, 204–206
Keer effect, 201
Kleiman rule, 203
$KNbO_3$, 203–204
Knock sensors, 246–247
Knoop hardness, 91

L

LaF_3 membrane electrode, 172
Lapping
  characteristics, 281–282
  mechanism, 297, 281–282
  model, 281
Laser, see also specific types of lasers
  amplification, glass for, 190–191
  Co_2 lasers, 288
  cutting, 291, 293–294
  drilling, 288, 290–291
  excimer, 299
  factors affecting, 286–287
  factory production processes, 289
  features, 285–288
  glass, 189–190
  machining, 300
  precision, 287

removal processing, 287–288
scribing, 293, 295–296
solid, principle, 200–201
types of lasers used, 286
Laser-assisted etching, 297–300
  device structure, 298
  feed speed, 299
  volume removal, 297–298
Lattice vector, reciprocal, 153
Lattice vibrations, energy, 151
Light emission, crystals, 202–203
Light modulation, 205–206
$LiNbO_3$, 203–204
Lineage, 27
Liquid ion exchange membrane electrode, 172
$LiTaO_3$, 203–204
  lapping, 281–282
Lodestone, 125

M

Machine tools for grinding ceramics, 277–279
Magnesia, 47–48
Magnetic properties, 125–128, see also Ferrite
  hysteresis magnetization curves, 127–128
Magnetism, origin, 126
Magnetite, 125
Maxwell-Eucken relation, 157
Mechanical capabilities, 233–234
Mechanical properties, 32–33, 83–84, 233, see also Strength
  elastic constants, 89–91
  flexural strength, 84–88
  fracture toughness, 91–93
  hardness, 90–91
  nonoxide high-strength ceramics, 234–239
  tensile strength, 88–89
  thermal shock resistance, 235
  zirconia, 239
Mechanochemical polishing, 283
Medical applications
  artificial joints, 217–220
  clinical, 217
  materials
    alumina, 216–217
    carbon, 214–216
    multicomponent, 217
Medical ceramics, types and strength, 211–212

Metallizing, 72–73
Mica ceramics, cutting, 266
Microhardness, 91
Microwave, ferrite
 applications, 139–142
 characteristics, 128–129
Miniaturization, 104–105
Mixing, body, 67–69
Modern ceramics, definition, 7
Molding, 65–66
 calendar roll method, 15–16
 classification by, 12–16
 doctor blade method, 14–15, 72–74
 extrusion, 13–14, 74–75
 injection molding, 13–14, 71–72
 isostatic pressing, 12–13, 69–71
 method types, 76–77
 slip casting, 13, 74
 vibratory solid casting, 13
Mold pressing, *see* Isostatic pressing
Multilayer ceramic circuit substrate, 108–109

N

Na-S battery, 177
NASICON, 121
Negative temperature coefficient
 thermistors, 110
Nernst equation, 119
New ceramics, definition, 7
Noncontact polishing, 284–285
Noncontamination machining, 285
Nonoxides
 boron nitride, 59, 61–63
 high-strength ceramics
  applications, 236–239
  development projects, 237, 239
  mechanical properties, 234–239
  target capabilities, 238
  thermal shock resistance, 235
 silicon carbide, *see* Silicon carbide
 silicon nitride, 56–59

O

Optical fibers, 193–199
 categorization, 194
 chalcogenide glass, 197–199
 graded-index rod, 196–197
 products, 196–197
 quartz glass group, 195

 refractive index distributions, 192–195
 relationship between losses and
  wavelengths, 195–196
 vidicon video imaging tube, 198–199
Optical properties, *see also* Chemical
 properties
 Faraday rotation glass, 191–192
 glass
  for acoustooptic elements, 192–193
  for laser amplification, 190–191
  lasers, 189–190
 optical fibers, see Optical fibers
 photochromic glass, 192–194
Oxide
 alumina, 31, 37–40
 catalyst, 184
 magnesia, 47–48
 zirconia, 40–43, 47
Oxygen sensor, 170–171
 automotive parts, 243–244

P

Permselective membranes, 169–170
Phonon–phonon collisions, 152–154
Phonons, 151
 mean free path, 154
Physical characteristics, 32–33
Pockels effect, 201
Polarization, 191
 within crystals, 201–202
Polishing, *see also* Lasers
 complex, 283
 cut-off methods, 302–307
 elastic emission machining method, 284
 electric discharge complex grinding,
  307–308
 electrical discharge machining, 307
 electrolytic complex grinding, 308
 hydration machining, 284–285
 ion beam machining, 308
 laser machining, 300
 mechanism, 282
 mechanochemical, 283
 noncontact, 284–285
 noncontamination machining, 285
 ultrasonic machining, 300–301
Polycrystalline ceramics
 thermal conductivity, 157–158
 thermal expansion coefficient, 161–163
Polycrystalline yttrium iron garnet,
 139–140

Polyethylene, ultrahigh molecular weight, 218–219
Porcelain, 104
Positive temperature coefficient thermistors, 111–113
Potential energy, versus interatomic distance, 149
Potentiometric sensors, 172–174
Potentiometry, 170
Powdered material, character, 28
Powders
 characteristics, 30–31
 impurity effects, 29–30
 selection, 67
 silicon carbide, 51–55
Precious-metal catalyst, 169, 184
Precision machining, 261–264, *see also* Cutting; Grinding
 lapping, 279, 281–282
 polishing, 282–285
 removal methods, 262
Pre-sintered ceramics, cutting, 264
Pressure sintering, 79–80
Production, 65–66, *see also* Molding; specific materials and processes
 body preparation, 66–69
 firing, 75, 78
 flowchart, 68
 future issues, 81
 integrated circuit packages, 73–74

Q

Quartz, 203–204

R

Reaction sintering, 78–79
Refractive index
 crystals, 201
 distributions, optical fibers, 192–195
Removal machining methods
 classification, 262
 laser processing, 287–288
 survey of problems, 263–264
Riser heaters, 246
Rockwell hardness, 90–91
Rubber pressing, *see* Isostatic pressing
Ruby crystal, 200, 202
 growth, 204
 laser drilling, 288, 291

S

Scribing, lasers, 293, 295–296
Secondary cell, 121
Selenium, use in copiers, 197–198
Semiconductors, 109–114
 boundary-layer capacitors, 113–114
 gas sensor, 169
 thermistors
  negative temperature coefficient, 110
  positive temperature coefficient, 111–113
 varistor, 113–114
Semihard magnetic ferrite, 136–139
 applications, 136
 cobalt-adsorbed, 137–138
 Fe_2O_3, 137
Shear strain, 224
Silica, reduction, 58
Silicification, carbon, 51
Silicon
 nitriding metallic, 57–58
 wafers
  resistless etching, 299
  scribing characteristics, 295–296
Silicon carbide, 48–56
 alpha, 49–51
 beta, 49, 51
 crystal forms, 49
 cutting, 266
 finer powder production, 49–51
 heat resistance at high temperatures, 110
 packing characteristics, 53, 55
 particle shape, 53–54
 particle-size distribution, 53
 powder characteristics, 51–55
 reaction sintering, 78
 sintered, 114
 sintering
  accelerator, 107
  characteristics, 55–56
 types, 49
 uses, 55–56
Silicon haloid, nitriding, 58
Silicon nitride, 56–59
 crystals, 56–57
 cutting, 266
 insulating properties, 107
 powder characteristics, 59–60
 production methods, 57–59
 reaction sintering, 78
 surface finishing process effect, 84, 86
 ultrasonic machining, 301
Silver halide, 193

Single-crystal materials
  cut-off methods, 302–303
  production technology, 22
Sintered body
  character, 28
  density changes, 55
Sintered ceramics, cutting, 265–266
Sintering
  accelerator, 107
  classification by, 17–19
  silicon carbide, 55–56
Slip casting, 13, 74
Slippage strength, 230
Sodium-sulfur cell, 121
Soft magnetic ferrite, 129–134
  applications, 133
  electrical power loss reduction, 132–133
  firing program, 130–131
  grain boundaries, 131
  limiting frequency, 129
  magnetic flux density, 132
  manganese–zirconia, 133
Solid electrolytes, 170
  fuel cell, 119–120
Solids
  thermal conductivity, 156–157
  thermal expansion, 159, 161
Solution method, 204–205
Special ceramics, definition, 7
Specific heat, 145–148
  constant-volume, 147
  curve according to Debye's model, 149
  Debye's theory, 147
  lattice vibrations, 146
  relation to thermal expansion coefficient, 150
$SrTiO_3$
  perovskite form, 117
  varistor, 114
Steam electrolysis, 183
Strength, 223–224
  bending tests, 230–231
  breaking strength, 226–230
  ceramics compared to metals, 236
  creep strength, 256
  defect diameter reduction, 241
  deformation, 224–226
    strength, 226–230
  fracture, 225
    mechanics tests, 231–232
  improvement, 239–241
  slippage, 228, 230

Stress intensity factor, 92
Strontium ferrite, 134
Structural ceramics, definition, 8
Substrates, 104–109
  fine-grained alumina, 107
  miniaturization, 104–105
  multilayered ceramic substrate, 108–109
  properties, 105–106
  required conditions, 105
Superconductors, 115–118
  applications, 116
  characteristics, 115–116
  critical temperatures, 115–116
  Josephson tunnel junction devices, 116–117
Surface-active ceramics, 214
Surface electrometer gas sensor, 169
Surface microflaw method, 92
Synthetic raw materials, 27–29, *see also* Oxides
  electrical characteristics, 36
  physical and mechanical characteristics, 32–33
  powders, 29–31
  thermal characteristics, 34–35

T

Technical ceramics
  definition, 7–8
  properties and uses, 24–25
Tensile strain, 224
Tensile strength, 88–89, 239–240
  hot-pressed PSZ, 243–244
  relationship with temperature, 255
Tensile stress, 228
Thermal barrier coatings, 253–254
Thermal characteristics, 34–35
Thermal conductivity, 95–96, 105, 235
  high, 158–160
  impurity effect, 158–159
  mean free path, 154
  phonons, 151, 156
  polycrystalline ceramics, 157–158
  porosity effect, 157–158
  solids, 156–157
  temperature dependence, 154–155
Thermal expansion, 148–151
  coefficient, 96, 150
    polycrystalline ceramics, 161–163
  cordierite crystals, 162–163
  low, 163–165
  solids, 159, 161

Index 353

Thermal properties, 93–94, 145
  ceramics that exploit, 155–165
  conductivity, *see* Thermal conductivity
  expansion, 148–151
  heat capacity, 94
  high-temperature applications, 234–235
  specific heat, 145–148
  thermal shock resistance, 96–99
Thermal shock
  resistance, 96–99, 235
    coefficient, 97–98
  temperature differential and breaking strength, 97–98
Thermal stress, polycrystalline, 163
Thermistor
  negative temperature coefficient, 110
  positive temperature coefficient, 111
Thermo sensors, automotive parts, 245–246
Three-dimensional structured conductors, 121–122
Tool material, hardness, 265
Turbine shrouds, 255

U

Ultrahigh molecular weight polyethylene, 218–219
Ultraminiaturization, 105
Ultrasonic machining, 300–301
Umklapp process, 153

V

Vanadium oxides, phase transistor, 114–115
Varicap, 113
Varistor, 113–114

Verdet's constant, 191
Verneuil method, 204
Vibratory solid casting, 13
Vickers hardness, 91
Vitreous carbon, implants, 212

W

Wiedemann-Franz law, 156–157

Y

Young's modulus, 224–226
Yttrium aluminum garnet, 202–204
  laser
    drilling, 290–291
    etching equipment, 297–298
    scribing, 295–296

Z

Zirconia, 40–43, 47
  characteristics and applications, 239
  cutting, 266
  dimorphous composition, 42–43
  elastic emission machining method, 284
  electrolyte oxygen sensor, 170
  monoclinic, 41
  oxygen sensors, 243
  partially stabilized
    strength, 240
    tensile strength, 243–244
  Y_2O_3, 43, 47
  stabilized, ionic conduction, 118–120
ZnO, sintered, 114